MACHINING
WITH ABRASIVES

MACHINING
WITH ABRASIVES

Richard L. McKee

VNR VAN NOSTRAND REINHOLD COMPANY
NEW YORK   CINCINNATI   TORONTO   LONDON   MELBOURNE

Library of Congress Catalog Card Number: 81-2991
ISBN: 0-442-25281-1

Manufactured in the United States of America

Published by Van Nostrand Reinhold Company
135 West 50th Street, New York, N.Y. 10020

Van Nostrand Reinhold Limited
1410 Birchmount Road
Scarborough, Ontario M1P 2E7, Canada

Van Nostrand Reinhold Australia Pty. Ltd.
17 Queen Street
Mitcham, Victoria 3132, Australia

Van Nostrand Reinhold Company Limited
Molly Millars Lane
Wokingham, Berkshire, England

15 14 13 12 11 10 9 8 7 6 5 4 2 1

Library of Congress Cataloging in Publication Data

McKee, Richard L.
  Machining with abrasives.

  Includes index
  1. Grinding and polishing. 2. Abrasives.
I. Title.
TJ1280.M35   621.9′2   81-2991
ISBN 0-442-25281-1   AACR2

Preface

This is a book about the machining of precision parts with abrasives, addressed primarily to people in industrial plants and shops, in universities and technical schools—to anyone, in fact, who can benefit by learning more about a group of processing operations which has the capability of increasing productivity, lowering costs, and creating better-quality products.

Grinding (the most common collective name for the majority of the processes) has been traditionally a finishing operation, as it is today for many precision parts. However, in foundries, where the primary purpose of the cleaning of castings is to remove the excess metal as economically and rapidly as possible, the established tool is an abrasive wheel and, to a lesser degree, an abrasive belt. This book is mostly about the in-between area of machining.

The basic idea of the book can be expressed in two ways. First, to assert that if abrasive tools—wheels and belts—are capable of fine precision finishes (as they have been for years), and if they are also capable of probably the heaviest kind of material removal (as they also are), then they should be capable of doing the intermediate machining that lies between these two extremes. Second, to declare that if a part has to be ground at some stage in its machining, then the possibility of doing all the machining with abrasives ought to be checked. The advantages of doing the whole machining sequence on one machine instead of two or more—with one setup rather than multiple setups, with a reduction in transportation, and often with less material to remove—ought to be apparent. Steel cutting tools must be taken off the machine and reground, often at some distance from the production floor. But grinding wheels are to a degree self-sharpening, for when sharpening (or "dressing," as it's called) is necessary, it can be done by the operator without removing the wheel from the machine. In fact, on high-production automatic operations, the wheel dressing is blended into the cycle without involving downtime. Abrasive belts and loose abrasives are simply used until they are worn out, at which time they are discarded. They have been termed the world's first throwaway tools.

Most of what has been written about abrasives has either dealt with the technicalities of the operations or discussed research. Things like the selection of the right grinding wheel or the chemical formulation are common in these books.

One need not be an abrasive engineer to understand that if a part can be completed from a raw blank—or from the solid, or whatever term is used for the part before machining—in one operation on one machine, it is probably possible to save money and time by doing it that way. Nor does it require an engineering background to understand that if the extra amount of stock needed for machining can be reduced, there will very likely be a saving in material.

So there is something to be said for a broad, if not necessarily deep, knowledge of abrasives and how they work, at a number of intermediate levels of industrial management and elsewhere. The foreman or first-level supervisor on the floor, as well as the floor process engineer, must be concerned with keeping his department's production levels up and his costs in line. He may quite possibly be aware of what today's grinding wheels and coated abrasive belts can do but have difficulties in getting his superiors to comprehend what he is talking or writing about. This book is intended primarily for his superiors.

The importance of the abrasives industry to industry in general far exceeds the size of its annual dollar volume, for the manufacture of abrasive tools ranks high in any nation's list of critical industries. But the industry is small in terms of dollars, and fragmented; there is no single trade association, for instance, for all the manufacturers of abrasives. And there are only a few—two or three—companies making a full line of abrasives, and probably fewer than that making both machines and abrasives. It is true that in the mass finishing segment of the industry, most suppliers provide all that is needed—the machines, the abrasives, and the other materials needed—but these are basically machine builders which buy the other products for resale under private labels.

A similar fragmentation exists in what has so far been written on the subject. For example, one book may have an extensive discussion of abrasive wheel surface grinders (for grinding flat surfaces) but never even hint at the fact that there are coated abrasive ("sandpaper") belt grinders which will do the same jobs at least as well, and sometimes even better. Or which may never mention that there are loose-grain machines certainly capable of improving flat surfaces and to some degree of generating them. Further, most discussions are organized on the basis of the way in which the machines are designed and not according—as they are in this book—to the basis of what they accomplish.

In this discussion there are some deviations from traditional terminology. One French term for a grinding machine is "machine á rectifier," which can be roughly translated as a machine to finish up or correct what some other process began. So the terms "grinding," "abrasive machining," and sometimes "abrasive cutting" have been used interchangeably, reflecting the author's opinion that the three are now synonymous, even though some individuals contend that abrasive machining is a broader term than grinding and that "abrasive cutting" ought to be limited to parting or sawing operations done with thin abrasive wheels.

Grinding, by whatever name it is called, now has the capability to be a broad-range and complete machining process in its own right, and should no longer be considered as limited just to secondary finishing of what has been begun by some other machining process.

But if there are milling cutters, as there are, and cutter-type machining with metal and carbide tools, why not abrasive cutting in a broader-than-traditional sense? When the term abrasive machining was introduced in the mid-1950s, it had tremendous value in its focusing of attention on a different approach to the use of grinding wheels. But now the time has come to recognize that whereas abrasive machining may be slightly more inclusive than traditional grinding, the three are essentially similar processes, and the terms can be considered synonymous.

The growth in the use of abrasives and, even more so, their increased potential have created a need for a book which will enable people without technical background in the subject to discuss it intelligently with those who do have some background. Such a book could also serve as an introduction and a reference to the subject, so it should be accurate without being overly technical. In keeping with that purpose, Chapter 1 is a basic introduction to the uses of abrasives that may be skimmed or omitted entirely by those with previous knowledge in the field.

Acknowledgments. Over the past three decades or so, the number of people with whom I have discussed abrasives and from whom I have learned what these tiny bits of mineral can do must be well up in the hundreds. However, I want to recognize specifically the help and encouragement of Bill Schleicher, who got me into this business to begin with, and of Cliff Duxbury, Chuck Nobis, and Doug Wachs. Without them this book could not have been written. But I should—and do—absolve them of any responsibility for any statements or conclusions contained in the text, which is entirely of my doing.

Richard L. McKee

Contents

MACHINING WITH ABRASIVES

1

The Basics of Machining and Abrasives

There are a few fundamental concepts and definitions about machining and abrasives that are useful in understanding this book. They are neither complex nor difficult, but without them it is harder to understand what follows. Anyone who has worked in a production or machine shop has probably acquired this information through experience; if so, this chapter can probably be skipped. However, in today's plants, experience in production jobs is not considered as important as it once was, so it is possible for individuals to go directly from college into a lower-management job with less than a casual exposure to what goes on in production departments. For such persons, among others, this book is intended. Managers need not know the details of what goes on, but it is helpful for them to be aware, for one instance, of what they are approving, without simply taking another's word for what is involved.

Machining is often regarded—and with reason—as a necessary evil. For that reason, considerable research has gone into the improvement of primary processes, such as casting, to achieve piece-parts that are closer to finished dimensions and configuration or geometry. (*Piece-part* is a term applied in many shops to units of product in process. It is so used in this book.)

Machining, broadly considered, is the process—or series of processes—by which a piece-part becomes a finished part. Sometimes it is more narrowly construed as the part of the process, toward the finishing end, involved in working to tolerances from the nominal dimension, like ±0.0002 inch. This means that the dimension can be over or under the nominal plus or minus two ten-thousandths of an inch, which is fairly close work with respect to commercial production grinding. Some parts, however, require only the removal of some flash (a ragged edge of excess material usually resulting from a forming operation) or burrs (ragged edges resulting from an operation like drilling). If the parts are small, they may be finished by holding them against a simple floor- or bench-mounted abrasive belt grinder; but if there is a large volume of piece-parts, the flash- or burr-removal job can be done more efficiently by loose bits of abrasive, in a barrel or vibratory finishing machine, or possibly by blasting.

CUTTING TOOLS

For years cutting tools have been pieces of hardened steel with either one or a few cutting edges, which could be mounted in a holder and moved across a piece-part under pressure to cut excess metal or other material from the piece-part. If the piece-part is mounted on centers and turned as the cutting tool is pressed against it, the operation is called turning, and the tool will usually have just one cutting edge. If the surface to be generated is flat, it is a milling job and the cutter will have perhaps a dozen or so teeth, and it will be rotated against the surface as the cutter moves across it—or as the piece-part moves under the cutter. A third type of cutter machine tool is the drill press, which makes holes in piece-parts. A fourth is the boring machine, which finishes and sometimes enlarges the holes made by a drill. There are many variations on these basic types, of course, but these are sufficient for the present discussion. Regardless of how hard and tough the steel is or how sharp the original edge, all steel cutting tools eventually have to be resharpened, a requirement that has led to the development of "throwaway" tools, usually made of tungsten carbide, a very hard material, which are held in steel holders, used until they are dull, and then discarded. At least that is the intent, and most shops follow such a practice. There are some shops that resharpen even throwaway tools, however, but this is considered of doubtful value.

Cutting tools are very efficient at removing excess material from piece-parts, but not so good for producing close dimensional surfaces or high-quality surface finishes. In fact, the first abrasive wheels and belts were developed because precision and surface-finish requirements had outstripped the capabilities of the cutting tool. Since that time over a century ago, even though cutting-tool capabilities have increased, so have the requirements for precision and finish. But there has probably been little change in their relative capabilities.

ABRASIVES

Abrasives can probably be best described as microminiature cutting tools, primarily small bits of hard mineral materials called grain, which can be cemented together (bonded is the trade term to be used throughout this book) into wheels of many shapes and sizes, cemented onto a backing of cloth or paper (sandpaper is the colloquial and coated the trade term), or used loose, without being held together in any way. These are the three principal forms. The largest grain is whatever will pass through a quarter-inch screen. The smallest is so fine that it floats in water, and is often sized by flotation, with each size depending on the amount of grain deposited as sediment within a given time.

When these grains of abrasive are bonded together in a wheel or adhered to a backing in a coated abrasive belt, they act like a milling cutter or turning tool, with perhaps millions of cutting teeth instead of the dozen or so characteristic or the milling cutter. (Microscopic studies have established that abrasive grains cut away the material they remove. They do not rub or wear away the material removed from the piece-part. In fact, a rubbing action would produce excessive heat, which is damaging to the abrasive cutting action and often to the work.)

Types of Abrasives

There are four or five types of mineral-like abrasives, depending on whether natural and manufactured diamonds are counted as one or two. Counting them as one is probably more practical, since diamond suppliers handle both.

The most feasible way of examining the types of abrasive is to group them by approximate price. The two cheaper and most-used abrasives are silicon carbide (Fig. 1), the first of the manufactured abrasives, and aluminum oxide (Fig. 2), which was developed only a few years afterward, around the beginning of the twentieth century. Tons of both types are made each year, although aluminum oxide is by far the larger of the two in volume. Both are made in different types of electrical furnaces. The per-pound cost of both is relatively low.

Natural diamond of industrial grade (Fig. 1–3)—any diamond not of gem quality—began to be used extensively as an abrasive shortly before World War II. Manufactured diamond (Fig. 1–4) was developed in the late fifties, and cubic boron nitride (CBN) about a decade later (Fig. 1–5). Both these abrasives are made under conditions of such extreme temperature and pressure that volume production has so far been impractical. The price of both has remained in the range of $3 to $4 per carat, or several thousand

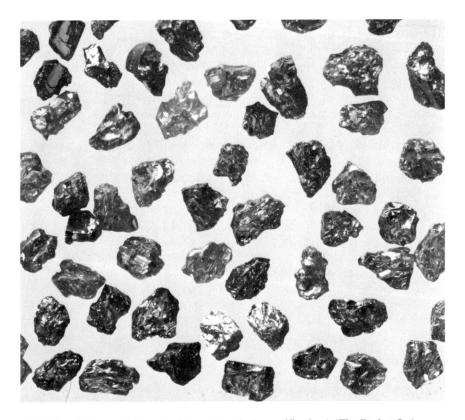

Fig. 1–1.   Silicon carbide grain. (Approximately 8 magnification.) (*The Exolon Co.*)

dollars per pound. (It is interesting to note that the first silicon carbide produced sold at about $880 per pound around 1900, which was then probably
equivalent to the diamond and CBN price today.)

   Of course, price is not the only differential; on some applications, diamond and CBN are so extremely efficient that this factor outweighs the
much greater abrasive cost, particularly on hard, difficult-to-grind
materials.

   Let's start with hardness. Diamond is the hardest substance known; CBN
is about two-thirds as hard; and silicon carbide and aluminum oxide,
about one-third as hard. On the Knoop scale, these are the relative ratings:

Diamond	7000
Cubic boron nitride	4700
Silicon carbide	2480
Aluminum oxide	2100

Fig. 1-2. Aluminum oxide grain. (Approximate 8 magnification.) (*The Exolon Co.*)

It has been observed that all abrasives are hard, but not all hard substances are abrasive. Indeed, there are several materials which are harder than either silicon carbide or aluminum oxide, but not so abrasive.

Other factors than relative hardness enter into the picture in determining which abrasives to use on what materials. Diamond is harder than CBN, and more effective on carbides, but less effective on steels. One of the disappointments in the use of diamond as an abrasive is that no one has yet come up with a coating or other variation that will enable it to grind steel efficiently. A similar situation exists with the conventional abrasives. Aluminum oxide is more effective on most steels and less so on nonferrous metals and nonmetallic substances than is silicon carbide. The best explanation of the problems of diamond and silicon carbide on steel is that there is a chemical reaction between the abrasive and the steels which in effect "melts" the abrasive and causes excessive wear.

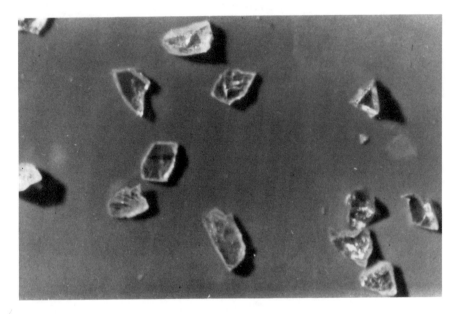

Fig. 1-3.   Natural diamond. (Perhaps 2 magnification.) (*General Electric Co.*)

Choice of Abrasives

With respect to the two lower-priced abrasives on the one hand and the two types of high-priced abrasives on the other, the choice is an engineering decision, because there is no significant difference in price. However, when it comes to a choice between, say, an aluminum oxide wheel costing a few dollars and a CBN wheel costing a few hundred dollars, price becomes a definite factor. Bear in mind, of course, that the higher-priced wheel may very well be so much more efficient and long-lasting that the per-unit abrasive price will be less than that for the low-priced wheel. Of course, the operator's experience and skill may also be an effective factor. As one shop owner once remarked, "I don't want anyone right off the street fooling around with a wheel that costs a couple of hundred bucks." He had a point. In fact, many shops feel that the price differential is still too great to allow CBN to be used on most production grinding operations, unless the hardness of the work material is such that aluminum oxide is not effective. Similar reasoning, of course, can be applied to diamond and silicon carbide. Diamond is the accepted abrasive for grinding tungsten carbides—in fact, for grinding most of the carbides. It is also the choice for cutting the "rumble strips" and other grooves in concrete highways.

Fig. 1-4. Enlarged metal bond saw diamond. (*General Electric Co.*)

Bonds for Wheels

For wheels the two principal bonds are vitrified (clay-type), which is used primarily with medium and fine abrasive grain for precision work; and resinoid (organic-type), which is used primarily with coarse grain for heavy stock-removal operations such as snagging castings. Rubber is also used as a bond for some specialized applications. There are many different formulations for each of these general types, and grinding-wheel manufacturers regularly come up with new ones, usually targeted at some particular market. And as their research engineers learn more about the effects of changes in formulation of the bond on wheel performance, it is sensible to conclude that bonds are improving.

Because grinding wheels and segments (Fig. 1-6) remove stock from piece-parts so readily, there is a strong tendency among operators to think of the wheel as being very tough and durable. Durable they certainly are,

Fig. 1-5. Uncoated cubic boron nitride. Enlarged. (*General Electric Co.*)

but tough, especially in vitrified bonds, they are not. The reason is not difficult to determine. The clays used as bonds in vitrified wheels are similar to those used for dishes; and the vitrifying process is similar to that used for dishes; so it is not surprising that the product is also similar to dishes. Resinoid wheels with a thermo-setting bond that is somewhat gummy before the wheel is baked are not as fragile, it is true; but neither are they so tough as to justify using a wheel that has been dropped or has received a similar physical shock. The risk of cracking is too great for a tool that when in use may be rotating at speeds of up to 3 miles per minute. (Vitrified wheels travel slower, at a little more than a mile per minute.)

Wheel Making

Most wheels are formed by pouring a mix of abrasive and bond into a round mold in which the mix is distributed as evenly as possible, inserting a

Fig. 1-6. Bonded abrasive segments. These are fitted into a holder, probably, in this style, to make a continuous flat grinding surface with the sides opposite to the marked sides. (*Bay State Abrasives, Dresser Industries.*)

top plunger into the mold, and compacting the mix under very high pressure. After vitrifying (at over 2000°F) or baking (resinoid, at 300 to 500°F) the wheel is finished to prescribed tolerances for dimensions and balance and is ready for shipment. Rubber-bonded wheels are handled differently, by a sort of cookie-baking procedure. The abrasive and rubber are kneaded together and rolled out in sheets of the required thickness when properly mixed. They are then cut or died out, just as with cookies, and baked. The scraps are kneaded, rolled, and cut out again, just as with cookies. Wheels can thus be made very thin—much thinner than is possible with the molding method, so many of the thinnest cutoff-type wheels are rubber-bonded.

Coated Abrasive Adhesives

Adhesives for coated abrasives function somewhat in the same way as bonds, in that they hold the abrasive grain in place. In the making process, a long web of paper or cloth about 52 inches wide is coated with a film of adhesive called the *make coat;* the adhesive is then applied (from the bottom up, drawn upward by electrostatic force); and after a drying period, a second film of adhesive called the *size coat* is applied. Then the long strip is dried again; and after a preliminary flexing operation to control the spacing and direction of breaks in the adhesive, it is ready to be cut, one way or another, into a finished product—belts, discs, strips, square sheets for incidental industrial and home shop use, sleeves, and other products.

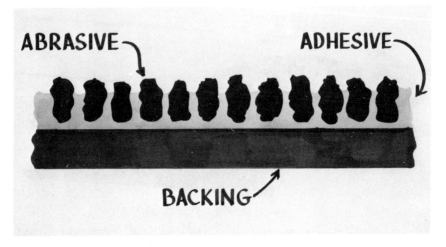

Fig. 1–7.   Enlarged side view of coated abrasive, with abrasive grain imbedded in adhesive on a backing. (*Hitchcock Publishing Co.*)

There are two principal adhesives—glue and resin—and the two coats of adhesive may be the same; or there may be resin over glue or glue over resin, depending on the final application intended. The advent of resin as a waterproof adhesive probably signaled the start of the use of abrasive belts for serious stock removal. Before that, when glue was the only adhesive used, belts could not be used with water or any other coolant, because the glue would dissolve and release the abrasive. But because of the development of waterproof coated abrasives, they are now used routinely with water or whatever other coolant is appropriate.

Coated Abrasive Backings

A word about backings. While they are generally considered to be paper and cloth, paper is actually limited to noncritical applications, such as the home-workshop market and very light duty applications. And the cloth backing used for belts must meet two somewhat conflicting objectives: it must be flexible enough to bend around the pulleys (contact wheels or rolls) on which belts (Fig. 1–8) are used and yet strong enough not to tear when pressure is applied—and pressure is obviously one of the key factors in any grinding. The really heavy, stiff cloths can be used as backing for abrasive discs, but the backing for belts must be more flexible. In fact, this one problem may well have been holding back the progress of belts into heavy stock removal applications.

Fig. 1-8. Coated abrasive belt. (*3M Company.*)

CLASSIFICATIONS OF GRINDERS

Not surprisingly, most of the grinding machines in function resemble milling machines and lathes or other machine tools. In fact, the first center-type cylindrical grinders of a century ago were lathes that had been modified to use a grinding wheel instead of the single-point cutting tool. And it would hardly be amiss to describe the most common types of surface grinder as milling machines with several thousands of teeth rather than a dozen or so.

It has been said that a grinding wheel or belt can do anything a cutting tool can do except drill a hole, and that is probably true. But once the hole is made, there are jig and internal grinders (Figs. 1-9, 1-10) that will finish the hole excellently.

Cylindrical Grinders

In a center-type cylindrical grinder (Fig. 1-11) the piece-part is supported on either end by centers and rotated for the stock-removal and finishing operation, duplicating the lathe. Either the piece-part or the wheel may traverse; the piece-part travels unless it is too large or too heavy to be moved readily, in which case it remains stationary and the grinder wheel travels. From a quality standpoint it is preferable to mount the wheel securely and to move the work. There is also less overall vibration.

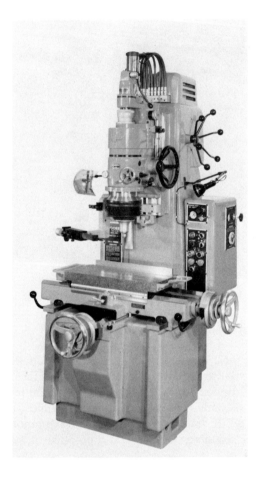

Fig. 1-9. Modern jig grinder for finishing regular or irregular holes. (*Moore Special Tool Co., Inc.*)

So far we've been considering the cylindrical grinder as having the wheel mounted at right angles with the axis of the work and the peripheral grinding face of the wheel square with the wheel sides and parallel to the work. This is the most common design. But there is also a design in which the wheel is mounted at or can be swiveled to a more acute angle with the work centerline, and its grinding surface is also dressed at the same angle, so that it is still machining in a plane parallel to the axis of the work. This is advantageous for some applications that will be discussed in detail in the chapter on cylindrical grinders.

It is also possible to design the machine so that either the wheel, as in

Fig. 1-10. Photo of internal grinder for grinding straight bores of cup-shaped piece-parts. Machine includes automatic loading-unloading and automatic sizing equipment. (*Cincinnati Milacron, Heald Machine Division.*)

Fig. 1-11. Center-type cylindrical grinder, traversing type. Note comparative width of wheel and length of shaft to be ground. (*Hitchcock Publishing Company*)

Fig. 1–12, or the work can swivel. This is convenient for grinding many piece-parts that are something other than straight, round shafts.

There are also applications where relatively short round piece-parts can be plunge-ground without traversing (Fig. 1–13). For this type of grinding, the length of the piece-part must be less than the width of the grinding face on the wheel. Similarly, there is a technique of cylindrical grinding on longer piece-parts which involves a series of overlapping plunge cuts, with the work indexed for each cut. For this, there is a final light traversing pass, simply for surface finish.

There are also special cylindrical grinders for grinding cams and crankshafts, and some very big ones for grinding steel mill rolls and related piece-parts. (Fig. 1–14)

Internal Grinders

Internal grinding is usually considered a form of cylindrical grinding, but it is technically considerably different and is done on an entirely different kind of machine. For one thing, the wheel for internal grinding must obviously be smaller than the hole in which it is being used, whereas the wheel for external cylindrical grinding can be of any reasonable size. This factor has more of an effect on machine design and the processes than might be thought at first.

Wheel Speed

To be an effective cutting tool, a grinding wheel must attain a minimum rate of speed to do the job well. The rate is expressed as *surface feet per minute* (usually abbreviated as *sfpm* or sometimes simply *fpm*). The rate,

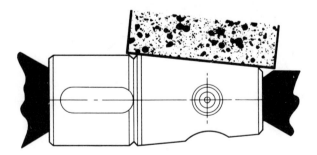

Fig. 1–12. Taper grinding on a center-type cylindrical grinder. Here the wheel has been swiveled, but the same effect results if the worktable is swiveled. (*Cincinnati Milacron.*)

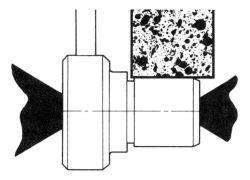

Figure 1-13. Sketch of center-type cylindrical plunge grinding, without any traverse. In plunge grinding, wheel (or belt) must be wider than length to be ground. (*Cincinnati Milacron.*)

Fig. 1-14. A large wheel like this form-dressed one operates efficiently at a low rpm. A 36-inch diameter vitrified wheel (normal safe maximum speed 6500 sfpm) could revolve at up to 700 rpm. In resinoid (9500 sfpm maximum), 1000 rpm. (*Hitchcock Publishing Co.*)

which is similar to miles per hour for an automobile, is the product of the wheel circumference, in feet, multiplied by the number of revolutions per minute (rpm). The minimum rate for a wheel is the lowest that is an effective cutting speed; the maximum is the highest safe speed, as determined by the wheel manufacturer.

Generally speaking, and with other factors equal, the higher the sfpm, the more abrasive grains are introduced into the cutting area, and hence the greater the stock removal. The second thing that makes sfpm important is safety. When employes are paid on the basis of their productivity, they have a powerful incentive to run the wheels as fast as they can, even to the extent of removing the speed controls or of making them inoperative. But the maximum safe speeds have been developed over a period of more than 50 years; they are frequently reviewed and adjusted to take into account the technical improvements in wheels and machines; and they are not arbitrary or capricious. Grinding wheel manufacturers have a definite economic stake in promoting safe wheel speeds. Product safety legislation and the resulting legal settlements for injury have carried the manufacturers' liability well beyond the first users to whom the wheels were sold—and sometimes regardless of the conditions under which the wheel was used or the speed at which it was operated.

From an operational and machine-design standpoint, the larger the wheel diameter, the fewer the number of revolutions it needs to attain an effective operating speed. At one extreme, the 60-inch-diameter wheels used in file grinding can attain an effective cutting rate at something between 400 and 425 rpm. At the other, a wheel only a fraction of an inch in diameter must reach a level of 100,000 rpm or more to be an effective cutting tool (Figs. 1–14, 1–15).

Both external and internal cylindrical grinding machines are more easily adapted to the production of formed piece-parts than are lathes. The distinction is that the grinding wheel—usually vitrified-bonded—can be readily altered to produce almost any form so long as its peaks and valleys run in the direction of the wheel's travel. This process of wheel-cutting surface alteration is termed *form dressing;* it can be accomplished by several different means, depending on the level of production, the precision required in the finished product, and other related factors.

In contrast to the lathe, which can generate forms usually by movement of the toolhead only and which can cut only one diameter at a time, a center-type cylindrical grinder can generate as many diameters as the width of its wheel surface (the thickness of the wheel) will accommodate. This can be a considerable production advantage, particularly for parts whose length is less than the thickness of the grinding wheel.

Fig. 1-15.   Internal grinding on a universal grinder needs high rpm to attain effective cutting speed. (*Hitchcock Publishing Co.*)

Surface Grinders

One might reasonably say that all grinders are surface grinders, but this term is usually reserved for grinders used to generate an essentially flat surface, which may have grooves or peaks and valleys running in the direction of wheel travel. Historically, the surface grinder can probably be regarded as an offshoot of the milling machine or the planer or the shaper, but probably nowhere in the history of machining have so many variations in machine design been developed for the accomplishment of one basic objective, the development of a flat surface by grinding.

Let's start with a few basic comparisons between cutting-tool machines and grinders that make flats. The first exerts tremendous pressure on the piece-part, such that it must be clamped or otherwise firmly held during the cutting operation. The grinding wheel exerts only light pressure, so that a magnetized worktable, usually called a magnetic chuck, provides sufficient holding power. (If the part is nonmagnetic, it can be held between steel blocks or bars; in fact, if there is sufficient production to justify the outlay, there are magnetic holders made particularly for the purpose of holding nonmagnetic parts.)

It is generally maintained that a cutting-tool machine will remove more stock than will a grinding-wheel machine, but this is no longer necessarily true. Grinding machines, particularly surface grinders, can often remove as much or more stock in a given time, and then, often by only an additional dressing to change the action of the wheel's grinding surface, finish the piece-parts to precise dimensions, surface finish, and flatness without removing the piece-part from the original workholder. Surface grinders for heavy stock removal are generally the larger, higher-powered machines.

The diversity of surface-grinder design was mentioned earlier. Here are some of the possibilities. The smallest surface grinders can be easily accommodated in an area about 3 feet square; the largest could easily be 50 or more feet long. The little machines can have motors putting out only a few horsepower; the biggest ones might have motors generating 300 or 400 horsepower—indeed, the major bar to greater power is the creation of a market for such a monster.

Grinding may be done on either the periphery or the side of the grinding wheel, and the piece-parts may reciprocate back and forth (traverse) and at the end of each traverse move either toward or away from the operator (cross-feed). Or the piece-parts may be held on a round, rotating magnetic chuck.

These possibilities immediately provide for the four basic types of surface grinders (Fig. 1–16). First and most numerous are those with a peripheral grinding wheel on a horizontal spindle, with the work held on a reciprocating magnetic chuck (Fig. 1–17). The second uses a side-grinding wheel held on a vertical spindle, with a rotating chuck (Fig. 1–18). The wheel's diameter is somewhat more than the radius of the chuck, so that it will grind all the piece-parts on the chuck. A third alternative is to combine a peripheral grinding wheel mounted on a horizontal spindle with a rotating chuck. This type produces a flat surface with a "scratch pattern" of concentric curved lines which is beneficial on many parts. Some machines of this design have chucks that can be tilted, which makes possible the grinding of either concave or convex dish-shaped parts. A fourth variation combines the reciprocating chuck with a side grinding wheel mounted on a vertical spindle. This design provides very high stock

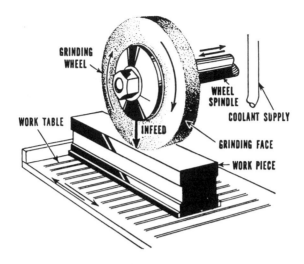

Horizontal spindle, reciprocating table grinding is best on jobs requiring great accuracy and good finish involving a straight pattern parallel to the traverse motion.

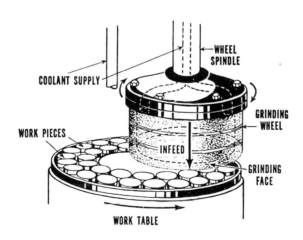

Rotary type vertical spindle grinding.

Fig. 1-16a. The two most-used types of surface grinders. (*Bay State Abrasives, Dresser Industries.*)

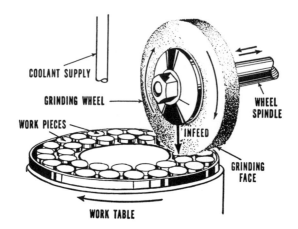

Horizontal spindle with rotary table. The scratch patterns (concentric curved lines) on this type of grinding can produce a perfect seal for a mating part — if the finish is fine enough.

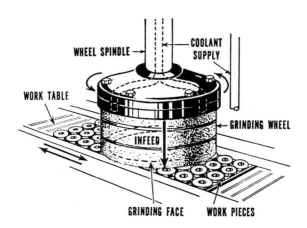

Typical vertical spindle reciprocating table surface grinding operation.

Fig. 16b.   Two widely used, but more specialized, surface grinders.

Fig. 1-17. Small horizontal-spindle, reciprocating-table surface grinder; operator stands at left, in front of the machine. (*DoALL Company.*)

removal and produces a cross-hatch finish which is considered superior for sliding bearing surfaces. And of course there can also be a side grinding wheel mounted on a horizontal spindle, for grinding a vertical surface of a piece-part. These latter types are specialized machines that have been created to fill a particular need. There are still other variations, but these are really extensions of one or another of these basic types.

If we consider workholding devices, we find that there are still other possibilities. A common one is two rotary chucks designed so that while one chuckload of piece-parts is being ground, another can be unloaded and loaded. Another, a grinder with an extra-long reciprocating chuck, so that while the batch of piece-parts at one end of the chuck is being ground, a batch at the other end of the chuck can be unloaded and set up for grinding. It's the same principle, of course.

The first two types are the most numerous and the ones that come to mind when surface grinding is mentioned. The horizontal-spindle reciprocating-table type is noted for its precision and, in larger sizes, by its form-

Fig. 1–18.   Vertical-spindle, rotating-table surface grinder, in action. (*Hitchcock Publishing Co.*)

grinding capabilities. The vertical-spindle rotating-chuck type is noted for high stock removal and high production.

Coated Abrasive Grinders

Both surface grinders and center-type cylindrical grinders can be built with coated abrasive belt heads rather than grinding wheels; in fact, it is not uncommon for a shop to modify a wheel grinder to use belts. Belts offer the possibility of combining the peripheral contact of a wheel with considerable width, so that cross feed of the part or the grinding tool may not be needed. Further, belts can also be used on machines which will include one or more roughing operations, intermediate operations, and finishing operations in one pass of the piece-part through a multiple-head machine (Fig. 1–19). With continuing improvement in belts and machines, this could be a fruitful field to explore for other than the surface-finishing operations where it has mostly been used.

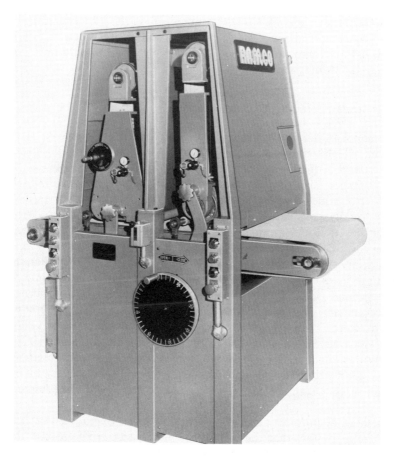

Fig. 1-19. Depending on the abrasive and the grit size of the belts used in this two-headed coated abrasive surfacing machine, it could be a grinder, a wood sander, or a polisher, with the new belts on the right-hand head for roughing, and the somewhat used belts on the left for finishing. (*Robert A. Martin Co., Inc.*)

Other Types of Grinding Machines

Though surface and center-type cylindrical grinders are most widely used and most competitive with cutting-tool machines, there are a number of other abrasive machines with which readers should be acquainted. It could well be that there are more different types of grinding machines—machines using abrasives, that is—than there are of any other broad classification. It seems as if whenever any plant needed a grinder to perform some special job, someone would come up with a design to do the job. There have been many such machines developed, often by only one company. But such specialized machines may remove considerable stock, in addition to their specialized finishing functions.

Other Stock-removing Grinders

As it happens, however, there are a couple of other standard types of grinders which remove substantial amounts of stock but do not have counterparts in the cutting-tool field. One is the centerless grinder (Figs. 1-20, 1-21), a high-production machine for producing essentially cylindrical parts, which because of its design and its general area of application is not usually considered with center-type cylindrical grinders. (In this chapter the traditional approach has been followed, but centerless grinders will be covered in more detail in Chapter 6 along with other types of cylindrical grinders.) The other machine is the double-disc grinder (Fig. 1-22), which produces two flat and parallel sides to parts at the same time. Parts to be ground on a double-disc machine are usually held between two bonded abrasive discs and ground by the sides of the discs at generally high rates of speed. Both centerless and double-disc machines are alike in that parts which have no protrusions to prevent throughfeeding—for instance, piston rings which are ground on the diameter by the centerless method and on the side by double-disc machines—can be processed at very high production rates. Details of double-disc grinding are covered in Chapter 5.

Centerless Grinders.   Centerless grinding is an operation that gets its name from the fact that the cylindrical piece-parts to be ground are not held between centers but rather between the outside diameters of two

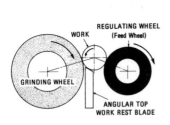 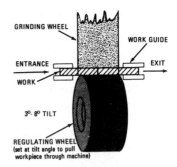

Fig. 1-20.   These two sketches show (left) a head-on view of centerless grinding and (right) a top view. This is through-feeding grinding, probably its most common use. (*Bay State Abrasives, Dresser Industries.*)

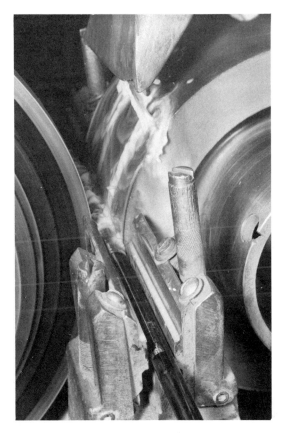

Fig. 1-21. Closeup of the exit side of a centerless grinder; likewise through-feed work. (*Hitchcock Publishing Co.*)

abrasive wheels and on top of a work rest. Its relation to center-type cylindrical grinding lies only in the geometry of the product; with respect to the grinding machines, there is very little relationship between the two processes.

Only one of the abrasive wheels does significant grinding; it is usually vitrified-bonded and rotates at standard speeds. The second, or regulating, wheel, which is usually rubber-bonded, runs at a much slower rate and acts as a brake to keep the piece-parts from spinning.

Centerless grinding is a high-production operation, particularly for round shafts or rods without shoulders, which can be through-fed from one side of the wheels and ejected from the other without any manual handling. The regulating wheel can be offset at a slight angle so that the

Fig. 1-22. Double-disc grinder into which operator is feeding automotive parts to be ground to length, on both ends, within close tolerances. Rotary feeder has pockets into which the parts fit. (*Hitchcock Publishing Co.*)

parts are pushed through as they are ground. Centerless grinding makes round parts rounder than probably any other process. (There are degrees of roundness.) Finishing piston pins and descaling and polishing steel rod stock are two other common examples of centerless applications.

Parts with shoulders or multiple diameters can also be ground, but the piece-parts must be retracted from between the wheels after they are finished. This is not so fast as through-feeding, but it has proved to be faster and overall more preferable to the other options for many such parts. And finally, since the diameter is generated from the center and there need be no concern about the location of the centers, less excess stock is needed for "rounding up" the piece-parts.

Double-Disc Grinders. Double-disc grinding is for flat surfaces what centerless is for cylinders—a high production precision operation (Fig. 1–22). Both processes use two abrasive tools; and in both the piece-parts are fed between the abrasive cutting surfaces. However, in double-disc work both sides are cutting, and the end result is two finished parallel surfaces. It is also possible to grind two similar parts positioned back to back between the discs, to take advantage of the machine's high-production potential. The usual situation is that two equal-area surfaces will be ground, which can be handily done with discs of the same specification. If the two surfaces to be finished have differing surface areas, some variation in the specifications for the discs must be made, but this is usually a minor problem.

For practical operations, the space between the discs is slightly wider at the entry end than it is at the exit end. Piece-parts must be confined by the feeder on all sides, and the feeder must of course be thinner than the piece-parts. Probably the simplest feeder is one which oscillates from a pivot in the machine so that it carries single piece-parts between the discs. This is the least-expensive, but also the least-efficient, feeder. It is primarily an arm with an opening of appropriate size and shape to carry the piece-parts one at a time. A more efficient type is the feed wheel, which is simply a disc of thin steel with openings around its outer area to carry piece-parts so that more than one can be in the cutting area at a time. Once the discs are set up, all the operator need to do is insert the parts into the feed wheel. After they are ground they fall out of the wheel near the bottom of its stroke to a tote box, a conveyor, or whatever other carrier is used to convey them to the next operation.

Abrasive Finishing Machines

The abrasive cutting machines that are generally comparable with cutting-tool machines usually have the range to be both stock-removal and finishing grinders. Finishing, of course, is an elastic term in that it depends on the requirements for the piece-parts, and some of these are not very stringent. However, there are a number of important abrasive machines which are for the most part limited to final finishing operations—processes which require only minimal stock removal but good to superior surface quality and/or flatness or roundness.

Surface finish will be discussed in more detail later. For the present, it is enough to say that nothing is ever absolutely smooth, perfectly parallel, exactly (many shops say dead) flat, or completely round. These are relative qualities; and even though perfection can be closely approached, it cannot be achieved. This has become more apparent as the instruments for measuring these qualities have been improved. Fortunately for the cost of machined parts, most of them do not need to approach perfection; they function adequately with less—sometimes with much less.

It is essential to be aware, also, that many surfaces need to be finished for the sake of appearance only (Fig. 1-23). Kitchen utensils are a common example. Thus the scratch pattern in the surface, mentioned previously, is most important, for it determines how light is reflected, and that in turn pretty much determines how the surface looks. However, most parts for industrial use must meet standards for surface quality (flatness or roundness)—qualities which can be measured.

Thus, a sufficiently magnified end view of a cross section of any machined surface will show that the surface is not dead flat but a series of hills and valleys, an enlargement of the scratch pattern. This need not be a defect. Even in bearing surfaces, there must be some minute crevices to hold oil to lubricate the surfaces; otherwise the oil simply slips out and is wasted. Naturally the surfaces cannot be too rough. If that occurs, wear is accelerated and something soon gives out.

Fig. 1-23. Buffing a frying pan on a cloth wheel whose face has been charged with a mixture of a very fine abrasive and a liquid carrier. Obviously an operation for appearance only. (*Hitchcock Publishing Co.*)

A word of caution regarding terminology is in order. Terms (for example honing) are used rather loosely in the finishing area. In this discussion the term *honing* will be used to describe an internal hole or bore finishing operation with bonded abrasive sticks or stones. Its counterpart for external finishing is called *superfinishing,* a term that originated with the developer of the process. In principle, it is simply external honing. *Lapping* includes a number of loose-grain flat finishing procedures using loose abrasive grain in a vehicle (usually some sort of oil), with the parts and the abrasive pressed between an upper (nonrotating) lap and a lower (rotating) lap. Any operation involving free-flowing abrasive grain comes under the general term of *mass finishing,* though there are subdivisions depending on the method used to agitate and move the grain.

The rate of movement of the abrasive grain in all these operations is relatively slow—slow, that is, in relation to the speed of a grinding wheel.

Honing. Honing, as it is generally considered in industrial usage, is an internal finishing operation done with abrasive stones mounted in a holder designed with a cone-shaped part that moves the stones outward in a concentric circle. In the operation, the outside surfaces of the vertically held stones (in vertical honing, that is) move in a combination rotation-reciprocation motion against the walls of the cylinder(s). The resulting cross-hatch pattern is an excellent surface to hold oil, which increases the lubricating capacity of the cylinder wall. And since either the holder or the piece-part (Fig. 1-24) is free to "float" (which simply means that it is not rigidly held in place), each honed hole is accurately rounded and in line with its own axis. The usual practice is to allow the hone to float, but in some small-diameter work the reverse is true.

The working surface of each honing stone is curved to an approximation of an arc of the circumference of the hole being finished, and each stone soon wears to the exact arc. There are other advantages to honing. The stones work first on the high areas of the cylinder, and because the outward pressure results from positive force from the cone rather than from spring loading, the stones do not move into low places until the circle of the stones enlarges to that diameter.

Superfinishing or Microstoning. Superfinishing is the application of similar principles to the exterior surface of a round or cylindrical rotating workpiece (Fig. 1-25). The unit resembles a lathe or cylindrical grinder in that the piece-part is held between centers while it is machined by a stone held by mild pressure against the outside diameter of the work and reciprocated back and forth along its length. The stone is dressed to the

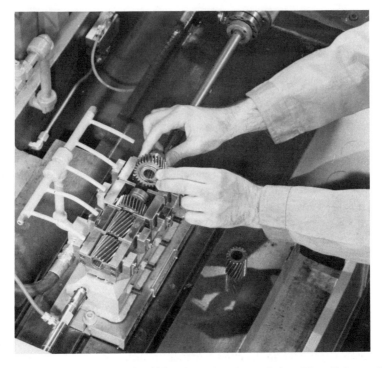

Fig. 1-24. Honing is unique in that either the tool or the workpiece "floats" (is not rigidly held) in operation. Hardened parts stacked horizontally, as above, can be honed to tolerances within 0.0002 inch at high production rates. (*Hitchcock Publishing Co.*)

same concave arc as the workpiece. Microstoning is an alternate term for this process.

Lapping. Lapping is an abrasive process for refining the surface finish and the geometrical accuracy of flat surfaces, and sometimes, in general usage, of cylindrical or spherical surfaces. Lapping is primarily a flat finishing operation. It removes minor defects left by preceding operations. The abrasive "tool" is usually a slurry of abrasive mixed with oil or soap and water, but sometimes a mixture of grease and abrasive. The "laps" are discs or plates of cast iron or some other metal which is softer than the piece-parts, so that the loose abrasive grains can embed themselves in the lap with enough of each grain protruding to cut the piece-parts. The piece-parts are usually free to move between the laps, but some applications use retainers whose function is similar to those used for double-disc grinding. With such a holder it is possible to lap cylindrical parts; the parts tend to

Fig. 1-25. This is actually external honing, although it is better known as either super-finishing or microstoning. The shaped bonded abrasive stone in the holder reciprocates left and right against the outside diameter of the rotating shaft, an action similar to honing, but external rather than internal. (*The Taft-Peirce Mfg. Co.*)

roll between the lap plates, while the holder serves merely to contain the motion (Fig. 1-26).

There is also a so-called lapping operation with one or two bonded abrasive laps and the work held in retainers between the two. One can be pardoned for considering this as merely a type of double-disc grinding with very fine grit discs, for as was said earlier, the terminology is not always consistent.

Mass Finishing. The link between lapping and mass finishing is that both use a slurry of abrasive grain and liquid. In mass finishing, however, there is relatively more abrasive than piece-parts, and the parts are immersed in a bath or mass of wet abrasive grain. Both parts and abrasive generally move freely, though sometimes the parts are fixtured or racked, because of weight which would otherwise cause the part to stay at the bottom of the tub when it should be surrounded by abrasive grain (the term in mass finishing is media). There is little indication that racking or fixturing is generally desirable; much of the time it is an unnecessary extra expense, but there are times when it must be done.

Barrel finishing (Fig. 1-27) or tumbling is the oldest of these methods, going back at least to the Middle Ages. At that time, pieces of armor or other metal pieces that needed to be shined up were put into some sand and

Fig. 1-26.   This machine could be used for either lapping or free-abrasive machining, depending on the hardness of the plates and the size of the abrasive grain used. (*Speedfam Corporation*)

possibly a little gravel in a barrel mounted on an axle on which it could be turned. Then the barrel was closed up and the metal pieces, with the sand and gravel mix, were tumbled together until the metal was sufficiently bright. The mixing action of the tumbling, particularly when it was end for end, was quite effective. Outside of making the barrel hexagonal instead of round, and improving the chemicals and abrasives used, the process hasn't changed a great deal since that time.

However, the development of the vibrating tub (Fig. 1-28) was a distinct step forward in mass finishing. For one thing, the container could be open instead of closed, so that parts in process could be examined at will, and the processing time could thus be readily adjusted to the work in process. For another, vibrating reduced the probability of part collision and consequent nicking and damage, because in the vibratory finisher the parts

Fig. 1-27. A barrel finishing machine with the guard down, ready to operate. The barrel is closed when operating, and when the load is finished, parts and media and compound are dumped into the inclined tray beneath the barrel. (*Rampe Mfg. Co.*)

automatically space themselves in the mass and proceed in a regular pattern. For a third, the vibrated media does an excellent job of removing internal burrs that cannot be removed by any other process. And finally, the open tub eliminates the possibility of any gas buildup. The action in the vibrator is continuous. Tubs can be rectangular, as well as round.

Spindle Finishing. A third variation of mass finishing is spindle finishing, with piece-parts held on fixtures extending outward from a central spindle. The mass of abrasive in which the parts are immersed is held in a tub which can be spun to make the abrasive flow around the parts. The fixturing eliminates any chance of part collision and damage, but the time required to attach and detach the parts is a drawback. Nonetheless, the greater speed of abrasive movement probably reduces the processing time.

As was stated in the introduction to this section, mass finishing removes very little stock. But in any plant which processes large quantities of odd-

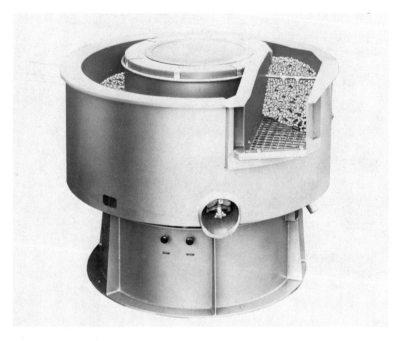

Fig. 1-28. Toroidal vibratory finisher, also known as a "finishing mill". Note the open top, together with the front media discharge door. Part and media mass unloading at the right. In operation, the unloading device is raised, so that parts and media move around the tub. (*SWECO, Inc., Finishing Equipment Division.*)

shaped parts of light to medium weight, where the objective is to produce good-looking parts free from burrs and surface flaws, some form of mass finishing can involve substantial cost savings.

Abrasive Action

To appreciate how abrasives work, in terms of their rate of stock removal and finish, it is essential to understand the role that such factors as grain size, wheel or belt speed, pressure on the abrasive tool, and the piece-part, and similar influences play. As is true throughout this chapter, these are generalizations that do have exceptions. This section deals with the general ideas; the exceptions, if they are significant, will be discussed somewhat further on.

One of the truisms of machining is that heavy stock removal and good surface finish are antithetical: you can do one or the other, but rarely both at the same time. This holds true for machining with abrasives. Coarse grain for stock removal and fine grain for finishing comes very close to being a universal generalization. It has, however, some exceptions. When the manager of a precision gage-manufacturing plant refers to roughing, he is assuredly not talking about the same type of operation as the manager of, say, a foundry.

If the abrasive is chemically compatible with the piece-part material and the material is of normal hardness, some of the factors that will determine the rate of stock removal and the quality of the surface finish are these: size of the grain, area of contact, wheel or belt speed, and pressure of the abrasive on the work.

Grain Size. Coarse grain removes more stock than does fine grain. However, once the piece-part material gets very hard, this ceases to be an advantage. In the hardest materials, there is no difference in the rate of stock removal, so one might as well use the finer grain to have more cutting in a given area. Moreover, the grain should not be so coarse as to leave scratches that the fine grain cannot later remove. Such an approach could ruin the surface finish and possibly leave metallurgical damage under the surface as well.

Area of Contact. Increasing the area of contact is another way of improving stock removal, primarily because there are more grains working on the piece-part surface at one time. As noted earlier, a vertical-spindle surface grinder, with a wheel cutting on its side surface, generates high rates of stock removal because there is virtually a flat-to-flat contact. Any peripheral contact is likely to be much reduced in area. Second in terms of abrasive-work contact is probably the internal grinder, with a convex arc of the wheel periphery contacting the concave arc of the hole. Third, but much less, is the contact between wheel and work on a horizontal-spindle grinder. Here, the arc of the wheel is in contact with a flat work surface. This makes a line contact, but one slightly greater than the arc to arc of a cylindrical wheel grinding a shaft. The least contact of all is the ball-grinding type (which has not yet been discussed), in which the arc of the wheel makes contact with a sphere and the area is thereby reduced to a point.

Wheel or Belt Speed. The role of wheel or belt speed is obvious. The greater the rate of speed, the more abrasive grains are introduced into the cutting area in a given period of time and the greater is the amount of stock removed. This holds true through all the recommended safe speeds, though it is interesting to note that experiments at higher speeds show some ranges where small increases in wheel speed do not increase stock removal.

However, as the rate goes still higher, stock removal resumes its upward trend.

Pressure. Pressure on the wheel or belt is another aid to higher stock removal, even though the bulk of machining with abrasives is considered as a light-pressure method. In billet grinding or foundry snagging, it has long been known that a strong and heavy man who can exert greater manual pressure on the wheel he is using can grind more billets or clean up more castings than can a lighter man who is not as strong. So by increasing pressure mechanically to several hundred pounds per square inch, steel mills and foundries have increased their grinding output tremendously.

The successful replacement of cutting tools by abrasive tools for heavy stock removal thus depends a great deal on the development of higher-speed—and consequently, higher-powered—machines that can exert greater pressure on the wheels or belts.

One of the factors holding back the increased use of belts in such situations is the development of belt backings that will be both flexible enough to move at high speeds around the contact rolls and wheels and at the same time strong enough to withstand considerable pressure without tearing—a point that will be discussed in more detail later. There are, of course, quite strong cloth backings used for abrasive discs, but such backings do not bend to any considerable degree. On the other hand, the area of contact between abrasive and piece-part surface in belt applications is almost without exception greater than that for wheels. Belts are characteristically wider; a 12-inch width is very common, the possibilities go up to 48 inches without splicing and even wider with splicing, though that increases belt cost significantly.

SUMMARY

No one has ever maintained that grinding—or abrasive machining, or cutting or finishing—is a simple process. Currently it is probably becoming a science, but with a generous infusion of art. The fact is there were—and maybe still are—many who call it akin to witchcraft. Some long-ago abrasive salesmen were suspected, with some basis in fact, of fostering such an attitude. Competition is keen, and comparative testing of abrasives to determine which of two or three different brands is superior is difficult and time-consuming, a procedure which can be justified only for high-production applications. However, if you have understood what has been written in this chapter, you should comprehend the following chapters and be able to understand what others are talking about when they discuss abrasives, and thus be able to make judgments with a better comprehension of the value of the proposed alternatives.

2
Abrasive Grain

Little bits of irregularly shaped, very hard minerals, less—usually much less—than a quarter of an inch thick, are the heart of any abrasive cutting system. All the parts of the grinding machine, which may total up to several tons, are designed and assembled for just one purpose: to make sure that the little bits of mineral cut the work material as precisely and as rapidly as is required.

It should not be surprising, therefore, that the major advances in machining with abrasives have come about through the development of new abrasives or of significant modifications of existing abrasives. We have thought in terms of five groups of abrasives—silicon carbide, aluminum oxide, natural diamond, manufactured diamond, and cubic boron nitride, but each has subtypes that have been developed to fill a range of needs. Aluminum oxide that is essentially pure, for example, is a fast-cutting but relatively fragile (friable is the term used in industry) abrasive. Some of the naturally occurring impurities actually make it a tougher, longer-lasting abrasive. And the addition of zirconia makes aluminum oxide still tougher and longer-lasting. Depending on the application, each has a place. Though wear life is a factor in choosing an abrasive, it is not necessarily the controlling factor; there are many applications in which a friable abrasive does a much better job.

DEVELOPMENT OF ABRASIVES

The development of silicon carbide and aluminum oxide around 1900 freed industry from its total dependence on natural abrasives of uncertain quality and composition, and gave abrasive cutting its first shove forward. The discovery of the abrasive qualities of diamond, about the time of World War II, facilitated the cutting of materials that the other abrasives just nibbled at; and thus, incidentally, provided a use for diamond boart, the scrap diamond left over from cutting gems, which had become a drug on the market. Diamond abrasive was an instant success, so much so that the excess was soon used up and the search for an artificial diamond, something that could be manufactured, was stimulated again. Natural diamond had to be rationed to applications for which it was most efficient rather than used for all the applications for which it was preferable to the established silicon carbide, primarily, and aluminum oxide.

Research efforts were rewarded during the late 1950s, when the General Electric Company announced that it had succeeded in producing authentic manufactured diamond. For 2 to 3 years, maybe more, teams of engineers from General Electric and a principal grinding-wheel manufacturer talked about the new development wherever they could find an engineering society meeting. It wasn't very long before DeBeers, the principal supplier of natural diamond, was able to produce a similar abrasive, and these two have been the principal suppliers since that time.

Cubic boron nitride (CBN) was developed a few years later, also by General Electric. CBN is manufactured by a technique similar to that which creates diamond, but is a substance not found naturally. It is useful in grinding hard steels which can be machined, though not always so effectively, by aluminum oxide.

DEVELOPMENT OF MACHINES

The development of machines has been a much more gradual process, with many small improvements. Even the practical application of centerless grinding, which came about 1920, was something that had been known for a number of years in principle: the first production of centerless grinders was preceded by a number of earlier, experimental machines which had demonstrated that centerless grinding was a workable process for producing cylindrical shapes.

TYPES OF ABRASIVE GRAIN

A point that was made earlier deserves reemphasis here. Even though we speak of four or five different types of abrasives—silicon carbide, aluminum oxide, natural diamond, manufactured diamond, and cubic boron nitride—and even though all of the abrasive grains of each type are basically similar, especially in terms of the kinds of material on which they are effective, considerable other differences exist. These have to be taken into account when different grains are compared. Color, for example, is a useful, but not infallible, means of distinguishing between grains of different subtypes. For example, the purest and most friable of the aluminum oxides is naturally white, so all the grain and wheel suppliers take great pains to maintain this color, even sometimes discarding wheels that are slightly off-color. Some suppliers even color their wheels as a means of product identification.

"Conventional" Abrasives

A number of factors make it practical to consider silicon carbide and aluminum oxide together. They are by a substantial margin the most-used abrasives; one furnace load of either, at any one of the manufacturers, probably weighs more than the entire annual production of either diamond or CBN, and quite possibly more than both put together. Between them, they do a generally good job of covering the range of industrial materials to be ground. And they are what most people think of when they consider abrasives or grinding wheels.

Silicon Carbide

It has never been made entirely clear what Dr. Edward G. Acheson was looking for in 1891 when he discovered silicon carbide, a substance that hadn't been known up to that time. What is known is that in one of his experiments he had mixed sand and coke, heated the mixture electrically to a point where in poking through the cooled mixture he found a few bits of a shiny black substance that was hard enough to scratch glass. What he thought he had found was something involving carbon and corundum, a natural abrasive much used at that time. So it was logical to call the new material Carborundum. He later organized a full-range abrasive manufacturing company with the same name. The company has been consistently a major manufacturer of the material from which it derives its name.

This abrasive is manufactured by the ton in troughlike furnaces (Fig. 2-1) for which typical dimensions could be 40 feet long, 10 feet wide, and 10 feet high when loaded. The sides are removable, and the ends are wired to conduct very heavy loads of electricity.

In the furnacing process pure coke is mixed with white silica sand, small amounts of salt and sawdust, and a generous amount of partly converted old mix. The furnace is partially loaded; a trough is scraped out in the top of the mix, between the two electrodes in the ends of the furnace, to hold pure graphite, which acts as an electrical conductor. After the furnace loading is finished, the current is turned on, and the mass eventually

Fig. 2-1. A bank of electric furnaces for producing silicon carbide which heat a loose mass of mostly coke and silica sand to over 4000°F and hold it for a run of a few days. The two in front are ready for unloading, with the imperfectly converted material scraped off. The fifth furnace back is still in process. (*The Exolon Co.*)

reaches a temperature of over 4000°F, at which point a large part of the load crystallizes in a sort of flattened tube. The duration of the cycle is a matter of days. When the conversion is complete, the furnace is cooled down and the sides are removed. Then the imperfectly converted mix, which is still grainy, is scraped off to expose the hollow core of crystalized silicon carbide. As mentioned above, the partly processed mix is reused in a later furnace load.

When the silicon carbide has been cleaned, the tube is broken into chunks (Fig. 2-2) for further reduction by crushing and rolling. For such a hard material, it is relatively easy to break up. At this stage the chunks of abrasive are quite pretty: regular grain is black or a somewhat iridescent dark blue; the purer type is dark green. Silicon carbide in chunks is also very abrasive; so that while souvenir pieces of the abrasive are attractive they can scratch any surface on which they are placed.

Subsequent treatment of silicon carbide involves a series of crushing and screening operations designed to reduce the size of the individual bits of grain and to separate them by size (Fig. 2-3). The sequence of operations

Fig. 2-2. A chunk of silicon carbide from early in the crushing cycle. Note its rather fragile appearance, in contrast to the aluminum oxide chunk in Fig. 2-5. (*The Exolon Co.*)

also modifies the shape of the grain—sharp and elongated, for coated abrasives; or blocky, for foundry and snagging applications; or somewhere in between, for general work.

Silicon carbide is the first of the successful manufactured abrasives to be developed, and in tonnage produced per year is second only to aluminum oxide. Apart from its historical significance, it has a definite place in the grinding and shaping of nonferrous metals, ceramics and other nonmetallic materials, and cast iron.

Silicon carbide is generally black, but some in a purer form is green. Wheels made with this grain are most often black because of the color of the grain.

Aluminum Oxide

Aluminum oxide is the most widely used abrasive in the world, primarily because it is used in grinding practically all kinds of ferrous metals. Its development occurred at about the same time as that of silicon carbide,

Fig. 2-3 Magnified about eight times, the end result is sized silicon carbide grain ready to be bonded into wheels or adhered to a backing. (*The Exolon Co.*)

though not quite in the same manner. Some time before 1900, aluminum oxide was identified as the abrasive element in emery, a widely used natural abrasive at that time. Its principal developer was Charles B. Jacobs, an engineer with the Ampere Electro-Chemical Company, who fused bauxite (an impure form of aluminum oxide named after the town of Les Baux, France, where the material was first quarried). After the fusing, he crushed the resulting dense mass into abrasive particles. This basic process is still being followed, though of course there have been many improvements in the details of the procedure.

The fusing process is simple to describe, but it was originally difficult to carry out, principally because of degree of heat involved. Pot-type furnaces (Fig. 2-4) of several tons capacity each are used. The heat is supplied by two electrodes. In the beginning, a small amount of bauxite (with some

Fig. 2-4. A battery of aluminum oxide furnaces, which are noticeably smaller than the silicon carbide "troughs." To heat these furnaces, electricity arcs between the tips of the two electrodes in each furnace. (The second barely shows through the smoke in the front pot.) Aluminum oxide fuses at over 4000°F. (*The Exolon Co.*)

minor additives) is loaded into the furnace and the electrodes are lowered into it. Sufficient current is sent through the electrodes to create an arc through the bauxite, which when heated to a temperature of something more than 4000°F becomes molten. As more bauxite is added, the electrodes are gradually raised, and the process continues until the pot is filled with molten bauxite. Originally the charge was left to cool in the original pots, but modern practice is to pour the molten material into cooling pots of construction similar to the furnace, where it can cool at leisure; the original pot is then refilled with another charge of alumina while the pot is still warm. This method reduces both the time and the electrical energy needed.

Fig. 2-5. A chunk of aluminum oxide. Looks much solider than the silicon carbide piece in Fig. 2-2, and it probably is. (*The Exolon Co.*)

After the "pig" has cooled, it is dumped out onto a floor for primary crushing with a "skull cracker" to break it into lumps (Fig. 2-5) for further jaw and roll crushing. After a series of crushing and sizing operations, plus possibly some roasting or heat-treating operations to increase the toughness of the grain, it is ready for fabrication into wheels or coated abrasive products (Fig. 2-6).

The abrasive characteristics of both types of abrasive are established during the furnacing and crushing operations, so very little that is done later affects those characteristics significantly. Note that most vitrified wheels carry the color of the grain—bluish-black or green for silicon carbide and white, gray, or tan for aluminum oxide. However, adding chromium oxide in small amounts will produce a pink wheel, and similar amounts of vanadium oxide create a green aluminum oxide, but these are for identification and marketing only. The additions have at most only a minor effect on the cutting ability of a wheel. Vitrified bonds are essentially colorless, so the color of the grain is generally the color of the wheel. Resinoid

Fig. 2-6.   Bits of aluminum oxide grain, at about 8X magnification; probably a random sample, but some look slivery enough to be used for coated abrasives. (*The Exolon Co.*)

bonds, particularly, are mostly black, so most resinoid wheels are black or very dark gray.

Types of Aluminum Oxide Grain

Aluminum oxide is essentially a tougher grain than is silicon carbide. Four types of gradations of toughness—or the lack of that quality, which is termed friability—are generally recognized. It might seem as though the toughest, longest-wearing grain would always be the best, but in practice this is not so. A grain that is too tough for an application will simply become dull and rub the piece-part surface, creating heat, which is the enemy of precision work. It is true, on the other hand, that too friable a grain will wear away very rapidly, thereby shortening wheel life, and this is not desirable either. But one need not stay in the middle all the time. Sometimes it is more desirable to have the wheel wear without overheating the piece-part; at other times for some applications where because of the wheel speed and principally the pressure involved, even the toughest grain wears down and is torn out of the wheel with ease. So there is a range of grain toughness suitable to a very wide spectrum of applications. And as it happens, the purer the raw materials, the more friable the grain.

Friable aluminum oxide, then, is made of relatively pure materials. It is white, practically always vitrified-bonded, and mostly used for tool grinding or similar precision work. In this range, the fact that the wheel may wear faster than normal is not a major consideration; toolroom grinding is by most standards not a production job. What is critical is that most tool steels are heat-sensitive, a criterion that far outweighs the consideration of the number of tools ground per wheel.

The next step is semifriable grain, an aluminum oxide whose titania content has been reduced to a little more than half that of the bauxite. It should be noted in passing that the sharpness of the fractured grain also affects its friability. The sharper the grain, the more friable it tends to be. Heating or roasting is another means of increasing toughness, if such should be desirable. In fact, through adjustment of the chemical content of the raw material, to some extent through modification of the fusing process, and through selection of the crushing and grain-treatment processes, the grain manufacturer has considerable flexibility in determining the final character of the grain, and can modify the product to take care of virtually any application within the broad range of the capability of the abrasive. Of course, this requires strict control over the grain-making processes by the manufacturer to ensure that for a given designation the same grain is delivered each time. It should also be mentioned that some crushing processes tend to produce a blocky shaped grain essentially for wheels, while others tend to produce grain for coated abrasives that is longer in relation to its thickness. The manufacturing process for wheels does not allow grain orientation, while that for coated abrasives does. Hence, grain for wheels must have several cutting points, moreso than that for belts.

Grain produced from bauxite with approximately its normal content of titania is considered "tough"; and if the molten abrasive is cooled quickly, the crystals formed in the fusing are small—hence the term "microcrystalline"—and the grain is considered extra tough. Another variation in the tough range is a grain consisting of about 25% zirconia and 75% alumina, cooled rapidly.

These tougher grains are used primarily in heavy stock-removal operations such as foundry snagging and steel-mill billet grinding. However, as one-step machining with abrasives—essentially producing finished parts from the "solid" or from rough parts entirely by abrasives—become more prevalent, it is entirely likely that these tougher grains will be more widely used for the stock removal needed in rough machining.

Physical Properties of Conventional Grain

In the discussions of silicon carbide and aluminum oxide, the principal physical property mentioned has been toughness or friability. Others which are important are grain size, hardness, shape (which is measured by bulk density), specific gravity, and crystal structure.

Grain size, oftentimes called grit size, is a tangible quality of all abrasives, both manufactured and natural. For the two conventional types, the accepted regulations concerning size are contained in two Department of Commerce publications, *Grading of Abrasive Grain for Grinding Wheels,* and *Grading of Abrasive Grain for Coated Abrasive Products*

Size. Originally, grain size was determined by the number of abrasive grains required to fill a linear inch. The method has never been described. However, with the adoption of United States standard sieves (with standard numbers of openings per square inch) in the early 1940s, the sizing process became much more accurate. And even though the standard sieves changed the sizes slightly, the original numbers have been retained. Their interpretation is easy: the larger the number, the smaller the grain.

Coarse sizes range from 4 and 8 to about 24; medium, from 30 and 46 to about 80; fine range, from 100 to 220; and powder sizes, from 240 to about 1000. These are not hard-and-fast groupings; the numbers are different for diamond and CBN, and the terms "fine" and "coarse" vary in meaning in different shops. For example, in a foundry, 36 grit would be regarded as a fine size, whereas in a precision machine shop, 60 grit or 80 grit would be coarse.

Abrasive grain of 240 grit and coarser is generally sized by passing it over a series of vibrating screens—usually several screens within each unit—which grow progressively finer. Grain smaller than 240 is sized by sedimentation or by centrifugal separation.

The minute size of abrasive grains, even the coarsest, is sometimes difficult to appreciate. For example, the sieve for a comparatively coarse 12 grit would theoretically have 144 openings per square inch, although the actual total would of course be somewhat less. For 240 grit, the number of holes per square inch would be something in excess of 55,000.

Another way of quantifying the size has been to estimate the number of grains per gram of powder. For silicon carbide, 240 grit, the estimate is 3,500,000. But for 600 grit, well into the powders range, the figure reaches 440,000,000.

Fortunately, the specifications do not call for grain to be 100 percent of the nominal size. Even though we talk of 60 or 100 grit, this represents a size with a small percentage of coarser grain and a somewhat larger percentage of smaller grain; because grain that is significantly coarser would scratch the surface too harshly for the rest of the grain to smooth out, leaving a low-quality surface. Finer grain presents no such problem, of course, but it simply does not have the stock-removal capability of coarser grain, which means that the finer grain does not pose much of a problem, but neither does it do much good. So the number given to any grain size means that the grain is primarily of that size, along with some coarser and some finer grain. Abrasive grain that totally includes the nominal size can be obtained, but the extra work required to refine it to that degree raises its price.

Hardness. All abrasives are significantly harder than the materials they cut, although that is not the only reason for the cutting. Hardness is usually defined as resistance to penetration by another material. In fact, the traditional method of comparing hardness of two materials has been to see what scratches what, as Acheson did with his first crystals of silicon carbide. In fact, one of the earliest comparisons, the Mohs scale, was established with 10 materials ranging from the softest, talc, as number 1, to diamond, number 10. Corundum, an impure natural aluminum oxide much used in the 1800s, is number 9; but there is a considerable gap between the last two numbers—9 and 10—of the sequence, enough to make the system of little other than historical value.

There are a number of hardness scales, mostly based on the principle of the depth made in the material being tested by a standard indenter under a standard pressure. However, because of the great range of hardness in materials, there is no standard that covers the entire range; and indeed, within a given test, it does not follow that a number twice as big means that one material is twice as hard as another. The two scales of most importance to users of abrasives are probably the Rockwell C scale for hardened steels (there are at least a half-dozen other Rockwell scales for a wide range of materials) and the Knoop hardness numbers, which are used for hard minerals like abrasives. The major abrasives on the Knoop scale line up like this:

Diamond	7000
Cubic boron nitride	4700
Silicon carbide	2480
Aluminum oxide	2100

On the Rockwell C scale (abbreviated R_c) a significant number of hard steels have numbers in the 45–50–60 range. For a very general comparison, since the relation is not a constant one, an R_c number could be multiplied by 100 for an equivalent Knoop number. This should be done with caution, however. For example CBN, at 4700 Knoop, is not absolutely 100 times as hard as a steel at 45–50 R_c. And there are factors other than relative hardness that determine the difficulty of grinding. Some relatively hard material can be ground quite readily; some comparatively soft materials cause all kinds of problems in grinding.

Grain shape is another important factor which is determined during the processing for the use for which a particular lot is intended. For grinding wheels there is no way of orienting grain in the wheel-making process, so the grain for wheels is blocky. For applications where high pressure is needed as in snagging, blocky grain withstands pressure better than does elongated grain. For all coated abrasives products, however, the making (manufacturing) process includes a step in which the grain can be oriented, so an elongated grain is preferable.

Tests for determining shape in abrasive grain follow the familiar pattern that a given volume of steel balls weighs more than the same volume of nails. Such tests are comparatively simple and readily adaptable to production quality-control requirements. Of course, microscopic comparison with standard samples is probably the best means, but is hardly applicable to production.

Other Physical Factors. Two other factors of interest are the specific gravity and the crystalline structure, which are established in the furnacing process and hence are not tested as routine procedures in the production sequence. The specific gravity of silicon carbide is about 3.2, and that of aluminum oxide, nearly 4.0. In comparison, water is, of course, 1.0, cast iron is 7.2, and gold, roughly 19.3. Generally, the larger the crystals, the more friable the grain. And the slower the cooling process for aluminum oxide, the larger the crystals. For very fine crystals the charge is cooled as quickly as possible, and the abrasive grain is fused in relatively small 1- or 2- ton pigs. Coarse crystalline material results from large 5- or 6- ton furnace loads, which are allowed to cool in the furnace shell.

Superabrasives

The term "superabrasive" has come into use as a group name for the two high-priced, high-performance abrasives; natural and manufactured diamond (Fig. 2–7) and cubic boron nitride (Fig. 2–8). There is no question about the superiority in performance, on hard-to-grind materials, of the two, nor about their cost. As it happens, both are manufactured by processes involving both extremely high temperatures and pressure. There are basically two major suppliers in the world.

Very little has actually been published about the details of the manufacturing process, and considering the whole situation, it is not likely that there will be any such publication for quite a while. There is really no need to know. Moreover, there are probably very few companies in the world with the know-how and the resources—plus the desire—to bring out a competitive product.

Fig. 2-7. Manufactured diamond, probably 40–50 mesh size at 60X magnification. (*General Electric Co.*)

Obviously, in a superabrasive wheel the most-expensive element is the abrasive, and in the specification for the wheels there is a spot for the thickness of the abrasive layer on the wheel core. Grain size numbers for the superabrasives are roughly the same as for the other abrasives, although the numbers are larger, indicating that the grains are not as large.

SUMMARY

Perhaps the most significant information that anyone who is not directly concerned with machining can glean from a discussion like this is that (1) abrasives, even the very expensive ones, are not really very costly in terms of unit cost, and (2) the tool (i.e., abrasive) cost is rarely the major expense

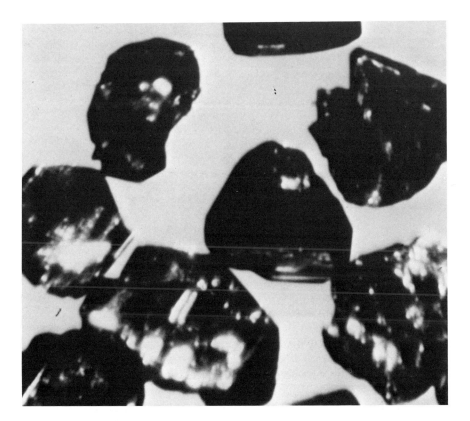

Fig. 2-8.   Manufactured cubic boron nitride (CBN), probably 60 grit—slightly finer than the diamond—at 100X magnification. (*General Electric Co.*)

item in the total machining cost. Direct labor time, setup time, transportation, tool-resharpening time, and the quality and consistency of the finished parts are all likely to weigh more heavily in any decision about machining than are the comparative costs of the tools involved.

3

Bonded Abrasive Products

Bonded abrasive products (Fig. 3-1) are often considered to be abrasive wheels or other shapes made of either silicon carbide or aluminum oxide grain cemented together with one of several bonds, particularly a glass-type inorganic bond known in the industry as vitrified, or an organic bond such as resin or rubber. Vitrified wheels are used for practically all precision work, resin (or resinoid) wheels are used for heavy stock removal in foundries and steel mills, or wherever the purpose is to grind off a lot of metal. The group also includes sticks, stones, segments, and other shapes made of the same grains and held together the same way. Bonded products may contain inserts of various kinds, either inserted during the molding operation or cemented in after the wheel is finished. The temperature at which the wheel is fired determines when the insert is placed.

Wheels and other products made of either diamond or CBN grain are also considered bonded abrasives, although the expense of these grains dictates making wheels with a thin layer of abrasive bonded around the rim of a plastic or metal core. These superabrasives are also made in other shapes, though rarely of solid abrasive.

Most abrasive grain that goes into a wheel made from conventional abrasives never cuts anything; it simply supports the cutting layers. It is the same grain, of course, but when the rpm of the machine will not revolve the wheel fast enough for cutting, the stub must be either discarded or moved to a faster machine, one fast enough to make the grain cut.

The three elements of a bonded abrasive wheel are the abrasive, the bond, and air. Wheels are not solid. Vitrified wheels are porous enough to allow coolant (sometimes called grinding fluid) to flow through a hollow spindle, through the bushing—if any—between the wheel and the spindle,

Fig. 3-1. An assortment of grains and bonded abrasive products. In front are mounted wheels and points. Behind them, three piles each of silicon carbide (left) and aluminum oxide (right); and next, a chunk of each kind of grain. Upper left are a straight cup and a flaring cup wheel, with two straight wheels behind them. In the center is a small straight wheel, with a couple of "plugs" for mounting on a heavy portable grinder. At top right are a couple of thin cutoff wheels. (*Hitchcock Publishing Co.*)

and out to the wheel-piece-part interface. This idea has never really caught on, even though it is an efficient way to deliver fluid to the interface, but it does demonstrate the porosity of vitrified grinding wheels. Resinoid and other organic-bonded wheels are not so porous because of the flow of the bond under heat.

Grinding wheels come in sizes ranging from little mounted points 1/8 inch in diameter and 1/4 inch thick or small slitting wheels perhaps a couple of inches in diameter and thin enough to slit the point in a pre-ball-point fountain pen to thick centerless wheels (Fig. 3-2) to huge wheels 5 feet in diameter and 12 inches thick (in one piece). Side-grinding "wheels" for grinding flat surfaces can be made of abrasive segments held in place on a wheel mount (Fig. 3-3). But the biggest grinding wheels of all are probably the pulp wheels (which convert logs into wood pulp) made of segments bolted into place around a central core.

Wheel sizes are usually stated in a standard pattern: diameter, thickness,

Fig. 3–2. The grinding wheel (left) in this centerless setup is one of the thickest molded. The regulating wheel is at the right, with the wide coolant nozzle near the grinding wheel and automatic gaging equipment between. (*Hitchcock Publishing Co.*)

hole size, in inches. For example a 12 × 1 × 1 wheel is 12 inches in diameter, 1 inch thick, and has a 1 inch hole. Inch marks may or may not be used. There are a few variations.

STANDARD WHEEL MARKINGS

Most grinding wheels made from conventional abrasives, excluding the very small ones, carry markings which indicate to those who know the code the type and size of the abrasive, the grade or hardness of the wheel (which is a different characteristic from the hardness of the abrasive itself), the structure (or degree of porosity), and the bond type. However, two points about standard wheel markings must first be understood.

Fig. 3-3. Bonded abrasive segments mounted in a vertical-spindle holder. The wide spacing between the segments makes for easy removal of swarf from grinding. (*Hitchcock Publishing Co.*)

Wheel Choice Limited

While there are theoretically several thousand different combinations of five factors—abrasive type and size, grade, structure and bond—for a given application only a few that warrant consideration. For example, suppose that a wheel is to be selected for precision cylindrical grinding of high-speed steel. Since the piece-parts are steel, and quantities are high, the abrasive will most likely be aluminum oxide, thus eliminating all the possible silicon carbide wheels. Every wheel manufacturer uses different aluminum oxides, but the field is narrowed considerably. There is thus a choice between two or three different grit sizes and about the same number of wheel grades and structures, though there might even be a "standard" structure, which would eliminate choice in structure. And finally, it would be a vitrified bond of some kind, so there might be a choice among three or four different vitrified bonds. But all the organic bonds have been eliminated. More important, the original thousands of choices have been pared down to a relative handful. An experienced abrasive engineer starting from this level may actually have to choose between only two wheels which are identical except for a difference in grade or grit size. If he is that close, he will likely ask for trial orders of the two wheels and make his choice on the outcome of the trial.

Symbols Standard, Not Wheels

The second caution about the meaning of the standard markings is that whereas the numbers and letters are standard, differences in manufacturing procedures with identical markings from two different suppliers may not be —and actually will probably not be—equally effective. Grit size, of course, is standard; it conforms to U.S. standards; but there are even different mixes of grit sizes in addition to the standard, as will be explained later. It takes some doing to determine which of the various kinds of types of abrasives, wheel grades, and bonds match up with similar designations from other suppliers. The suppliers, of course, have data showing how their products compare with other suppliers' wheels, but these data are generally not available. However, in any plant using large quantities of wheels, it should not be too difficult, over a period of time, to accumulate such information from trials or actual use. The specifications for the wheels that are being used on any particular application are a good starting place for any salesman wishing to displace the wheel being used with one of his own.

Anyone directly involved with selection of abrasives should of course be familiar with the details of the standard marking system (Fig. 3–4). And anybody who is at all involved with machining should understand these details enough to follow what they say about a wheel. The marking system is not all that complicated; much of it follows a logical pattern that can be readily picked up. In this section only silicon carbide and aluminum oxide wheels will be covered; markings for diamond and CBN wheels are similar, but different enough to warrant separate treatment.

Abrasive Type

A full-line wheel supplier might easily offer as many as a dozen different kinds of aluminum oxides and half-a-dozen silicon carbides. These range all the way from a friable white to an extremely tough alumina-zirconia alloy. In a symbol or marking such as the following,

$$29\underline{A} \quad 60 \quad K \quad 5 \quad V9$$

The A in the first position indicates that this is an aluminum oxide. (This is, by the way, an actual recommendation by one of the wheel-making companies for cylindrical grinding of high-speed steel. Like all such recommen-

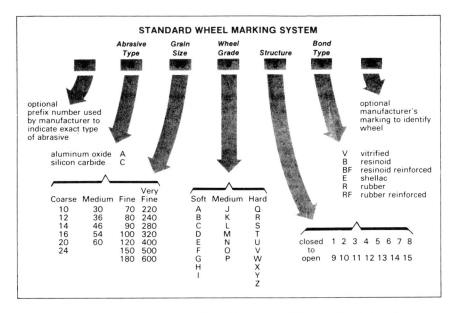

Fig. 3–4. A chart that shows the standard marking system with a complete listing of the symbols for the "conventional" abrasives, silicon carbide and aluminum oxide. (*ICS-Intext, Inc.*)

dations, it is intended as a starting point for determining the final recommendation for an actual application on a particular machine in a given location grinding a given material.)

In this company's listing, 29A is listed as one of two next-to-toughest aluminum oxides. The 29, of course, is the company's identification of a particular kind of aluminum oxide; it is an optional marking.

Grit or Grain Size

29A   <u>60</u>   K   5   V9

The 60 in the second position of the marking symbol tells the size of the grain in the wheel. As has already been mentioned, the larger the number, the smaller the grit size. This (see Figure 3–4) is considered about the smallest of the medium-size group. Of course the grouping as indicated in this table is rather arbitrary; in a precision grinding-shop working to tolerances in millionths, 60 grit would be considered quite coarse.

The conventional axiom about grit size is: coarse grain for stock removal; fine grain for finish. It does have exceptions. For one, if the piece-part material is exceptionally hard, coarse grain will not remove significantly more stock than fine. So it is better practice to use a finer grain than normal, because there will be more small chips removed thereby. The smaller the grain size, the more cutting grains there are in a given area of wheel surface. The second exception is that it is occasionally desirable to mix two or sometimes more sizes of grain in a wheel. Such a mixture is called a combination grain; it is indicated by a digit added to the basic size number. This could cause a problem, because there are fine-grit sizes with three digits, like 600, in the basic number. The solution is simple. If the size number in the marking symbol ends in a zero, then a grit size in the powder range is indicated. If, however, the number ends in some other digit, say 3, as in 603, then the grain is a basic medium 60 grit in a 3 combination, a much larger grain. Finally, it should be reiterated that any size designation is a nominal one, though the final screening would produce a breakdown something like this: no grain more than two sizes larger; a maximum of 25 percent one size larger; 52.5 percent or more of the total either of nominal size or one size smaller; and a maximum of 3 percent finer than two sizes smaller. Most of the grain of any nominal size is thus either that size or one size smaller.

It is obviously impossible to monitor the size of all the grain processed on a production basis, but all the abrasive grain producers maintain continuous random sampling of their product, and any lot not meeting size standards is rerouted for additional screening.

Wheel Grade

Grade could easily be considered the most baffling of the wheel characteristics to understand, probably because it is. By definition, it is a measure of the strength with which the bonding material holds the abrasive grain in the wheel. One might think that it would be best to use a bond that would hold grain indefinitely, but if this were the case the grain would become dull, burnishing the piece-part rather than cutting it, and heating it up in the process. Ideally, if the wheel is right on grade, grains will be shucked out as soon as they reach a certain degree of dullness; but in practice this rarely happens, so the wheel must periodically be dressed (i.e., have dull grain removed by passing a diamond or other type dressing tool across its face as it rotates) to remove spent grain. On the other hand, if the bond is not strong (i.e., hard) enough, grain will be released prematurely, while it is still capable of cutting, and the wheel will wear out more rapidly than it should, (Fig. 3-5).

Thus wheel grade depends on the strength of the bond holding the grain, which in turn depends on the percentage of bond in the mix. The more bond, the stronger the holding power, and the harder the grade. (Fig. 3-5.)

Weak holding power

Medium holding power

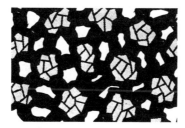

Strong holding power

Fig. 3-5. Three levels of grade are illustrated above, ranging from soft (top) through medium to hard (bottom). The bond shows as increasingly thicker "posts" connecting the abrasive grains. (*Bay State Abrasives, Dresser Industries.*)

"Hard" and "soft" in reference to the grade of abrasive wheels probably arose from the early days of wheel manufacture, when someone with a screwdriver or an ice pick graded the wheels as hard or soft according to the indentation that could be made in the wheel with the tool.

Grade is indicated by a letter of the alphabet in the third position of the marking symbol:

29A   60   <u>K</u>   5   V9

with the range nominally from A (extremely soft) to Z (very hard). In practice, E grade is about as soft a wheel as any company makes; and grades harder than Z are manufactured for applications involving heavy stock removal and very high pressures, like foundry snagging and billet grinding for scale removal. K grade would be considered toward the soft end of the medium grades. As with size and stock removal, there is also a trade axiom: "Hard wheel grades for grinding soft materials; soft grades for grinding hard materials."

Grade may well be the most important single factor in a grinding wheel's performance; it is certainly the most intangible. (A number of wheel manufacturers have introduced intermediate grades such as J+, which is purported to be midway between J and K.) Grade is based on the percent of bond and abrasive grain in the wheel mix. It is checked during and after the manufacturing process by either electronic or mechanical means, and the final marking which goes on the product is based in varying degrees on all these determinations.

But there is another complicating factor beyond the control of the wheel manufacturers. Grinding conditions such as wheel speed or work speed, or the hardness, softness, or "stringiness" of the piece-part material may make the wheel "act soft" or "act hard" depending on whether, under those conditions, the grain is held less tightly or more tightly than is normal for the nominal grade.

Research and experience in the area of operating variability of grade has been summarized in what is known as the "grain depth of cut" theory; anything that increases grain depth of cut also increases the force tending to tear the grain from the wheel, and thus makes the wheel appear softer. Any condition which decreases the grain depth of cut has the opposite effect.

Here are some of the conditions which make a wheel appear softer than its nominal grade. (These are based on the assumption that other conditions remain constant.)

1. Increase of work speed
2. Decrease of wheel speed
3. Reduced wheel diameter
4. Reduced work diameter

The extent to which these conditions can be manipulated in practice varies a great deal; there may frequently be questions about whether they can be altered on any one production batch to make a difference. On most grinding machines the work speed is easier to change than is the wheel speed; wheel speed is often constant. Furthermore, most machines are built for a certain diameter, or perhaps a range of diameters, of grinding wheels, although it is true that as the wheel wears, its diameter decreases. For any one production lot, the work diameter is fixed, but if another lot has a substantial difference in diameter (this variation applies only to cylindrical grinding, of course), there could be reason to consider a different grinding wheel. Piece-parts for surface grinding either reciprocate on a square or rectangular magnetic chuck, or rotate on a round chuck under the grinding wheel. The reciprocating speed can be controlled, as can the rotation, and in the latter case, there is some variation depending on whether the particular piece-part is toward the rim or the center of the chuck. In fact, in loading the chuck, a blank space is left in the center, because any piece-parts placed there would not move fast enough for the grinding action to be effective.

Choosing the optimum grade in a grinding wheel specification requires consideration of the material being ground, the condition and location of the grinder, and most important of all, the area of contact between the wheel and the work.

The hard material–soft wheel, soft material–hard wheel relationship was mentioned earlier; its justification is as follows. On very hard materials, neither aluminum oxide nor silicon carbide is going to make very deep scratches or cuts on the material, so it is preferable to use a wheel that wears away and frequently exposes fresh, sharp grain. Of course, if one can justify the cost of either a diamond wheel or a cubic boron nitride wheel, then that is the way to go. On soft materials, a harder wheel of conventional abrasive can dig in to remove stock without a significant problem of wheel wear; the abrasive grains stay sharp in spite of their digging in.

Soft-grade wheels can be very effective when used on machines that are comparatively free of vibration, whether the vibration comes from within the grinder (from worn spindles, for example) or from without, from failure to insulate the grinder from railroad or other traffic-caused vibration or from location on an upper floor of a building. Where there is vibration, grinding wheels must be definitely harder in grade—other things being equal, of course.

For many, the most important factor in grade selection is the area of contact between the grinding wheel and the piece-part. At first thought it might be concluded that all such areas are very small, and indeed they are, except for side grinding wheels generating flat surfaces on piece-parts on a surface grinder, and for internal grinding wheels grinding the inside surfaces of holes. The first exception involves a flat-to-flat relationship, with resulting low unit pressure; and such applications require the softest wheels, like grades E and F, and occasionally harder ones, like J, K, or L. The most-common application of this kind is on a vertical-spindle surface grinder, where the grinding is done either with the flat side(s) of a cylinder wheel (a side-grinding wheel with a hole almost as big as the wheel diameter) or with the flat side of a cup wheel, or with abrasive segments in a holder to form a wheel-like tool. All these are possibilities for flat surface grinding.

Next smaller—much, much smaller—in area of wheel-work contact is internal grinding, where the contact is between the external arc of the periphery of the grinding wheel and the somewhat larger internal arc of the work. Wheels for internal grinding are rarely softer than grade J, and range up to about grade M. These would be considered medium-hard wheels.

Peripheral surface grinding, where the contact is between the outside arc of the wheel and a flat surface, probably has a little less contact area.

Peripheral cylindrical grinding, with contact between the large arc of the wheel and the smaller external arc of the workpiece surface, takes somewhat harder wheels, although still mostly in the medium range.

Structure

Structure in the standard wheel marking is indicated by a number (usually 1 to 12) immediately following the grade letter of the marking. Structure is considered to be in the fourth position in the symbol. It indicates the grain spacing, or grain density of the wheel, with the large numbers representing the more open spacings. The voids, incidentally, allow the coolant to be fed, as it occasionally is, through the wheel. The voids also provide chip clearance, so that the bits of metal removed from the piece-part surface may be thrown off by the wheel. If a wheel is too dense, that is, if the voids are too small, the bits of metal can be retained in the wheel face (a condition called loading) and eventually these bits will reduce the cutting efficiency of the wheel. And in view of the fact that a vitrified grinding wheel revolves at something more than a mile a minute, the spaces in the wheel create a breeze with significant cooling action, which causes such a wheel's action to be known as "cool cutting." A normal structure is illustrated in Fig. 3-6.

Fig. 3-6. Normal structure in a wheel. Probably in the range of 6 to 8 in the standard marking. (*Hitchcock Publishing Co.*)

Where exceptionally wide spacing (Fig. 3–7), described by a number 13, 14, or 15, is desired in a vitrified wheel, to the mix is added something like ground walnut shells, which burn out in the vitrifying or firing cycle, leaving behind small extra voids, because the pressed and dried wheel is close to its finished size by the time it is ready for firing.

In the sample specification we have been using (29A–60–K–_5_–V9), the 5 structure is relatively dense. It is one that is frequently used in centerless and cylindrical grinding and in snagging.

Since both structure and grade are determined largely by the relative amounts of abrasive and bond in the mix, there has been a trend toward elimination of the structure symbol entirely. Structure itself is not being disregarded; rather, there has been a trend toward considering that in relation to the other elements of the marking, particularly bond and grain size, there is a "standard" structure which is best for that combination. The standard structure would of course vary with different combinations of the other parts of the marking, so there would not be one across-the-board standard.

Changing structure is not the only way of dealing with wheel loading, nor, for that matter, is grain spacing (structure) the only way of providing for chip disposal. In wet grinding, the coolant tends to help prevent lodging of chips; indeed, some grinders are equipped so that a jet of coolant is directed toward some point in the wheel for the sole purpose of dislodging chips. In dry grinding it is sometimes possible to impregnate the wheel with

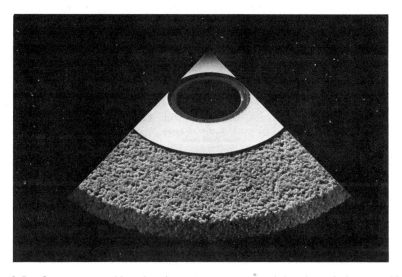

Fig. 3–7. Open structure. Note that there are more pores and that the grain is more widely spaced than in Fig. 3–6. (*Hitchcock Publishing Co.*)

a substance to fill the voids and reduce loading, or to coat the grinding face of the wheel with stick tallow or something similar.

Bond Types

If one single factor in the standard marking system could be singled out as most important, it would be the bond, which is indicated by a letter in the fifth position. And since most manufacturers use more than one of each type of bond, there is usually a second symbol, in what is sometimes called the sixth position, to indicate which particular formulation has been included. In the wheel marking which has been used as an example (29A–60–K–5–*V9*), V9 designates the bond—a vitrified bond for which the manufacturer's formulation is 9.

The symbols used are as follows:

Vitrified	V	Shellac	E
Resinoid	B	Rubber	R
Resinoid reinforced	BF	Rubber reinforced	RF

The derivation of the symbols for vitrified and rubber bonds are obvious. The other two are a little more obscure. Shellac was formerly known as an "elastic" bond, hence the E. And whereas today resinoid is used to a much greater extent than was rubber, rubber was used first. The earliest resinoid-type bond was Bakelite, which accounts for the B.

In general terms, the bond determines the maximum safe speed for a grinding wheel. Vitrified bonds are the slowest. (The maximum is usually considered to be 6,500 sfpm—a little more than a mile a minute.) Resinoid wheels for years had a maximum of 9,500 sfpm; but in some wheel shapes, and particularly when the bond is reinforced, they may go as high in industrial usage as 16,000 sfpm, which is about 3 miles per minute. Of course all grinding machines are designed with wheel guards to protect the operator at the safe speeds (and for that matter even at much higher speeds).

Wheel bond also determines the method of manufacture and thus has an influence on wheel design. For example, resinoid-bonded wheels, which are cured at about 4000°F, can be reinforced with plastic webbing or with molded-in iron rings. But this design is not possible with vitrified wheels, for which the firing temperature is above 2000°F. Both vitrified and resinoid wheels are pressed in molds, a process that limits the minimum thickness; rubber wheels are rolled out like cookie dough and then cut to size, again like cookies, and so they are limited in thinness only by grain size. The thinnest abrasive wheels are rubber-bonded.

Bond type has also had an influence on the different applications of wheels. Early in this century vitrified was the dominant bond. It was used for practically everything. With the development of resinoid bonds, however, the possibility of higher speeds and consequent higher stock-removal rates led to the displacement of vitrified for work like foundry snagging and billet grinding. At the present, this is the situation: Because vitrified-bonded wheels are inert to all grinding fluids and can be readily shaped for form grinding, they are the choice, for a couple of reasons, for precision grinding of all kinds. Resinoid wheels are the choice for high-speed, high-pressure applications involving maximum stock removal. Most regulating wheels in centerless machining are rubber-bonded. Cutoff wheels, sometimes called abrasive saws, are almost all rubber- or resinoid-bonded.

Advantages of Standard Markings

The adoption of the standard wheel-marking system was a considerable step forward in the use of bonded abrasives, even though it contains factors—grade is probably the prime example—that are difficult to measure. And it has not become sufficiently standardized to eliminate the differences caused by variations in processing by different wheel suppliers or by their differing interpretations of the requirements. Each wheel supplier knows, of course, how to convert competitive wheel specifications. He knows, for an example, that grade K is equivalent to another's grade L, or perhaps J. Volume users of bonded abrasive wheels can also approximate this information if they want to do so, but for users it is more often an impression they get rather than a demonstrable similarity.

The standard wheel marking, however, is the shorthand of the industry for bonded abrasives made with conventional abrasives, and some knowledge of it is necessary for anyone who wants or needs to understand these tools. It also helps to explain why the choice of the best wheel for a particular operation is still somewhat a matter of trial, and error, a procedure that wheel suppliers generally welcome. It may also help one understand less-than-satisfactory performance.

But even at less than peak performance, an abrasive wheel that is only approximately the correct specification can save time and money on many applications; and when everything goes together, the bottom line on a change to grinding from some other machining process can look very good.

Manufacturing Process—Vitrified

Most vitrified wheels are pressed, a process that is generally divided into the operations of mixing, molding, drying, firing or vitrifying, finishing, and inspection.

Mixing.   In mixing (Fig. 3–8), measured amounts of carefully prepared feldspars or clays (the bond) and abrasive grain are carefully weighed and thoroughly mixed in power mixers. Small amounts of other materials, like the powdered walnut shells mentioned earlier, may be added to provide the finished product with the desired characteristics. The mix is moistened with water or another temporary binder to make the wheel stick together after it is pressed.

Molding.   During molding (Fig. 3–9), the mixture of bond and abrasive is uniformly distributed in a steel mold and then compressed (Fig. 3–10) in a powerful hydraulic press to form a wheel, or sometimes a block, somewhat

Fig. 3–8.   One of the early steps in wheel making is to mix grain and other ingredients—some from overhead bins, some from smaller container—in the huge mixing bowl at the left. (*Bay State Abrasives, Dresser Industries.*)

Fig. 3-9.   Distributing the mix in the mold. The wheel molder tries to get the mix as level as possible to produce a wheel in better balance. He has, of course, weighed out the mix to the exact weight specified by the process engineer. (*Bay State Abrasives, Dresser Industries.*)

larger than its finished size. The amount of pressure varies according to the structure desired. The molded product is placed on ceramic batts to dry.

In this green stage, the ware (product) must be handled very carefully because it is fragile. Sticks and stones, which are essentially rectangular shapes, are often cut in the green stage rather than being molded individually. And small wheels can be cut out of green slabs by fly cutters on a drill press. Vitrified product is molded straight, without recesses or relieved sides; but if such sides are required, they are generally "shaved" after the piece is dried, on a device resembling a potter's wheel. Some vitrified bonds cannot be treated this way however, so if relief or a recessed side is required, it must be hogged out after the wheel is vitrified, a much more difficult procedure.

Fig. 3-10.   Once the mold has been closed, the next step is to press the wheel to shape. The 2000-ton press closes top and bottom plates simultaneously to achieve uniform density through the wheel. (*Bay State Abrasives, Dresser Industries.*)

Vitrifying.   Firing or vitrifying was once done in brick periodic kilns into which ware had to be carried, piece by piece, until the load was complete. Then the door was cemented shut and the heat was turned on. Now considerable firing is done in tunnel kilns, where the ware on batts is placed on a moving belt for a ride through the kiln. When the ware comes out the other end, it is vitrified. This is mostly for small wheels, sticks and stones. Larger wheels are fired in a bell kiln, a kiln in which the ware is stacked out in the open on a base (Fig. 3-11) to a prescribed height, after which the bell or cover is lowered over the stack and secured. Then the heat is turned on. At the end of the run the bell is removed, and the ware can be easily unloaded. Vitrifying is done at temperatures in a range of about 2000 to

Fig. 3-11.   These vitrified wheels—"ware" is the floor term—have been bedded for firing under the huge cover at upper right. This type of kiln eases the job of building the load within the kiln. (*Bay State Abrasives, Dresser Industries.*)

2300°F. It is a process similar to that for making china dishes—and the end product has resemblances to dishes, including easy cracking.

Once the firing is complete, only the size and the shape of the wheel can be changed. Its quality and cutting effectiveness have been established.

Finishing.   In the finishing operations the wheels are trimmed to finished size and shape with hardened steel cutters, other abrasive wheels, or diamond cutters. With proper application, the steel cutters do a surprisingly good job of removing abrasive from the wheel. The arbor holes may simply be reamed to size, or they may be bushed with lead, babbitt, or another material. Any metal insert, such as a mandrel for mounted wheels, threaded inserts for discs, or cylinder wheels bolted onto backing plates rather than on a spindle as most wheels are, is cemented into place.

Other finishing and testing operations include balancing (Fig. 3-12), which is similar to that for automobile tires and wheels; speed testing for wheels over 6 inches in diameter; grading and marking; plus a final inspec-

Fig. 3-12. Like automobile tires, wheels must be balanced by adding weight at the right places to ensure smooth rotation at high speeds. (*Bay State Abrasives, Dresser Industries.*)

tion (Fig. 3–13) to ensure that the wheel is of the required shape, size, and specifications, and that it is not chipped nor cracked.

Speed Testing. In the speed test, wheels are run without load at 150 percent of the established safe operating speed for their bond and size. Research by the wheel-making companies and through the Grinding Wheel Institute has established that this level is safe for a sound wheel under load at the maximum safe speed.

The market share for vitrified-bonded wheels with conventional abrasives is probably around 50 percent. There has been a trend toward its replacement with resinoid bond, but some resinoid bonds react unfavorably with some coolants, while vitrified bonds are inert to all coolants.

Manufacturing Process—Resinoid

The manufacture of resinoid-bonded wheels differs from that for vitrified in the mixing and in the firing or curing operations. Otherwise there is little difference.

Fig. 3-13. There are many quality control checks during wheel manufacture. After the wheel has been trimmed to specified thickness, this inspector "mikes" the wheel to make sure specifications have been met. (*Bay State Abrasives, Dresser Industries.*)

Mixing and molding resinoid wheels requires air conditioning. In mixing, thermo-setting synthetic resins are mixed, in powder or liquid form, with abrasive grain and a plasticizer to make the mix moldable. Resinoid mix tends to be a little bit gummy, in contrast to vitrified mix, which is at most slightly moist and could generally be considered as a dry mix. Molds are similar to those for vitrified wheels, with the exception that the wheels can be molded to shape. And molding to shape is done for two reasons: it can be done without danger of cracking during curing, and resinoid wheels cannot be easily shaped in the green state. Molding to shape also conserves on mix, which is not, however, a major factor.

After the wheel is hydraulically pressed to process size—and frequently to shape—in a mold similar to that for vitrified wheels, it is cured at a temperature of 300 to 400°F for a period of 12 hours to about 4 days, depending on its size. During this operation the mix softens first and then hardens as the oven reaches curing temperature. The bond retains its final hardness after cooling. The remainder of the making process is similar to that for vitrified wheels.

Resinoid wheels account for approximately 35 to 40 percent of all grinding wheels used today. There has been a continuing trend over several years away from vitrified and shellac bonds toward resinoid for high-speed rough operations like foundry snagging, portable grinding operations, cutting off, and, rather surprisingly, roll grinding steel mill rolls, which is presently one of the most precise of all grinding operations. The major reason for this swing to resinoid is probably because it is almost certainly the strongest of the bonds; wheels made with it can operate at higher speeds than those made with other bonds; and they can probably take rougher treatment than any others. This does not mean, of course, any relaxation in the safety requirement that a wheel be checked out if it has been dropped or bumped.

Historically, the problem of resinoid bond for wet grinding has been its tendency to soften and wear excessively when used with coolants. This has been common knowledge over a long time, and the problem has been attacked on three fronts—coolant research that develops new formulations that are kinder to resinoid bonds, grain coatings which resist coolants, and, of course, the development of resins that are more coolant-resistant.

Manufacturing Process—Rubber

Rubber bond is pure rubber, either natural or synthetic. (The mixing, or kneading, process by which it is combined with the grain was described earlier.) Sulfur is added as a vulcanizing agent. After the mixing operation, small batches are passed and repassed through calender rolls until the sheet reaches the specified thickness; then wheels like cookies are cut out of the sheets, to a specified diameter and hole size. Next the wheels are vulcanized in molds under pressure in ovens at about 300 to 350°F. Finishing and inspection are similar to the operations for other wheels.

The wheels are specially made for specific uses, such as wet cutting-off, for which they produce nearly burr-free cuts. (As has been noted, the manufacturing process permits the making of extremely thin wheels.) Rubber wheels are also used as regulating wheels in centerless grinding: as the second abrasive wheel in a two-wheel operation to keep the work from spinning at the same rate as the grinding wheel; and as finishing wheels on ball-bearing races where a high finish is required. Rubber wheels constitute somewhat less than 10 percent of the total market.

Manufacturing Process—Shellac

Shellac is a natural organic bond the use of which goes back a long way; today, however, shellac-bonded products make up only a very minor share of the total market. Inasmuch as the bond appears to provide a burnishing action on the work, shellac-bonded wheels are used to produce high finishes on camshafts, rolls, and cutlery.

To make these wheels involves mixing abrasive grain with shellac in a steam-heated mixer which thoroughly coats the grain with the bond. Wheels in the cut-off range of 1/8 inch thick or thinner are molded to exact size in heated steel molds. Thicker wheels are hot-pressed in steel molds. After the pressing, the wheels are set in quartz sand and baked for a few hours at about 300°F. Finishing is standard.

DIAMOND AND CBN WHEEL MARKINGS

Both diamond and cubic boron nitride wheels, because of the expense of the abrasive, are molded in a thin (1/32 to 1/4 inch) layer around a core, in contrast to conventional grinding wheels which are all abrasive. The marking system (Fig. 3-14) is similar to that for conventional wheels but with significant differences, that will be indicated in the text which follows.

The degree of standardization among diamond wheel makers is not the same as that of manufacturers of conventional abrasive wheels, even though a few of the bigger companies manufacture both types. Hence, in-

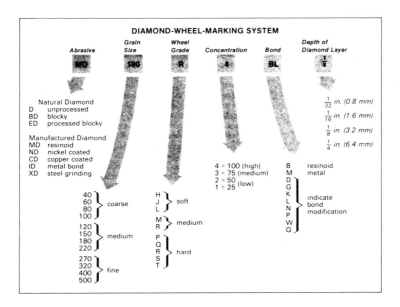

Fig. 3-14. A chart for a diamond wheel-marking system. Concentration, in the fourth position, replaces structure of the other system, whereas "depth of diamond layer" is new. (*Intext, Inc. Used by permission.*)

asmuch as the discussion that follows is a general explanation of the diamond (and CBN) markings, there are differences among the products of individual companies that make it difficult to make comparisons without some interpretation of each specification involved.

Abrasive Type

Starting from the left of the marking, the first position indicates the general type of abrasive: D for natural unprocessed diamond, and B for cubic boron nitride. Each may be modified by a preceding letter to indicate some processing or perhaps the metallic coating. Synthetic or manufactured diamond is always indicated by a two-letter symbol for example, MD.

Grain or Grit Size

The numbers and the sizes they indicate are substantially the same as those for aluminum oxide or silicon carbide. The major difference is that for diamond and CBN the coarse sizes are much finer than those for the others. For example, one diamond wheel maker lists coarse sizes as anything between 40 and 100; medium, 120 and 220; and fine, 270 and 500. Some size symbols indicate a range, as 100 to 120.

Wheel Grade

Grade, a letter in the third position, has essentially the same significance as in the conventional wheel marking system, although the range of grades is not as wide. The manufacturer just mentioned lists a range of grades from H on the soft end to T on the hard side. Some indication of the trend comes from the listing of "hard" grades P, Q, R, S and T. Diamond wheel grades tend to be on the hard side, but this has no relation at all to the actual hardness of the diamond abrasive.

Diamond (or CBN) Concentration

Diamond concentration is a term peculiar to the superabrasives, and it is indicated by a numeral—4, 3, 2, or 1—in the fourth position of the marking. Since it is determined by the volume of abrasive in the wheel, it does bear a resemblance to the wheel-structure symbol in the conventional wheel specification, but it is much more important because of its bearing on the price of the wheel. The number signifies the weight of diamond per cubic inch of the layer of diamond and bond (matrix) which makes up the grinding surface of the wheel. A 4 is the symbol used for what is termed 100 concentration, which is 72 carats of diamond per cubic inch of matrix. A 3 or a 2 indicates, in that order, 75 or 50 concentration, or 54 or 36 carats of diamond per cubic inch of matrix, respectively. It may be assumed that the fewer diamonds there are per cubic inch, the farther they are apart, hence the analogy to wheel structure.

Once this is understood, it is relatively easy to calculate the comparative weights of diamond in two different wheels by simply reading the specifications. All the necessary figures are stated in inches—wheel diameter and width, and thickness or depth of the diamond layer (the final element in the specification). The calculation is as follows: circumference × width × depth of diamond × 72 (for 100 concentration—or 54 for 75 concentration or 36 for 50 concentration). The result is the number of carats of diamond in each wheel. This is not a determination of which is best of the three; the most economical percentage of diamond may very well be different for any two applications.

Bond Type

Bond type is indicated by a letter of the alphabet in the fifth position, and any bond modification in the sixth; they are usually linked, as, for example, BL, a resinoid bond of the subtype indicated by the manufacturer by an L. The bond symbols are the same as are those for conventional wheels—B for resinoid, V for vitrified, and, a new one, M for metal-bonded. The modifiers vary with the manufacturer; they are not standard.

Depth of Diamond Layer

The depth or thickness of the diamond layer, as noted above, is the last element in the symbol, in the seventh position. The range is from 1/32 at the thinnest, to 1/16, 1/8, and at the thick end, 1/4. These are actual dimensions, not symbols or a code. The usefulness of this element of the symbol was explained above.

Core Material

Some manufacturers of diamond wheels add an eighth element to the wheel marking to indicate the kind of core used, but this is not a universal practice. If it is used, however, it will be accompanied by an explanation of its formulation. However, core material is generally determined by the manufacturer.

GRINDING-WHEEL SHAPES AND SIZES

If the description of a grinding wheel were confined only to its formulation, which has just been discussed, it would be incomplete. It is also essential to know the wheel dimensions diameter, thickness, and hole or arbor size—plus the dimensions and location of any recesses or relieved areas in the wheel.

The need for a code differentiating the various shapes of wheels was recognized early in the twentieth century, probably before the recognition that some kind of standardization in the expression of formulation was needed. The code was started appropriately enough with what is called a type 1, or straight, wheel, which is defined as one having a diameter, a thickness, and a hole. Each shape is designated as a "type" followed by a numeral from 1 to 30, although some numbers have been needed for a while, but later eliminated. Wheels are basically divided into those for peripheral grinding and others for wall or side grinding, although there is *one* which is considered safe for grinding on either the side or the periphery. The grinding face of each type is indicated, and it is not considered safe practice to grind on any other face.

Peripheral Grinding Wheels

The basic peripheral grinding wheel is the type 1 straight wheel, with three dimensions—diameter, thickness, and hole, the order in which the dimensions are listed. This is also by far the most numerous group. The smallest standard wheel, one for internal grinding, is $1/4 \times 1/4 \times 3/32$, though a 6 $\times$ 1/8 $\times$ 5/8 cutoff wheel is not very big. The largest one made in one piece is a 60 $\times$ 12 $\times$ 12 wheel for cutlery grinding. (Numerals are used for both shape designation—type 1, type 5, etc., and for dimensions. Dimensions are always expressed in inches, and in a set order: diameter, thickness, hole, as 60 $\times$ 12 $\times$ 12 above.) There are also some wheels for centerless grinding which are 22 or 24 inches thick to increase the cutting range of the wheels as the piece-parts rotate between; because in centerless grinding, the width of the wheels of the grinder determines the length of time that the piece-part will be ground (assuming a steady feed rate, of course). These wheels are made from two slightly angled halves cemented together. With a thickness of more than 12 inches it is difficult to maintain uniform pressure, and uniformity, within the mold. The hole of a type 1 wheel is usually much less than half the wheel diameter, although in some cylindrical grinding wheels the dimensions may be 20 inches $\times$ 2 $\times$ 12, or 30 inches or 36 inches $\times$ 3 inches or 4 inches $\times$ 20 inches. Wheel mounts are either 12 inches or 20 inches in diameter on this type machine, as indicated by the last numeral of each dimension, the hole size.

Other common types of peripheral wheels (Fig. 3–15) are type 5, which is recessed on one side, and type 7, which is recessed on two sides. (These are abbreviated rec. 1/s or rec. 2/s, respectively.) And there are other combinations which are either recessed or relieved (tapered inward) on either one side or both sides. The reason for a recess or relief is that it enables the user to make a desirable combination of productive capacity (the wide grinding face of the wheel) with a shorter and consequently more rigid spindle (since the thickness at the hole is less than the thickness at the periphery.)

Side- or Rim-Grinding Wheels

This group of wheels is designed for grinding on the flat side (Fig. 3–15), so the side or rim is usually narrow (except for type 27, which has a flat grinding face); but since the wheel—used mostly for portable grinding—is held at approximately a 15° angle to the work (Fig. 3–16), the effect is still that of a narrower grinding face. The exception to this is the bonded abrasive disc, which is a flat-sided circular piece with several countersunk holes scattered in a pattern across its face so that the disc may be bolted to a

TYPE 1 — STRAIGHT

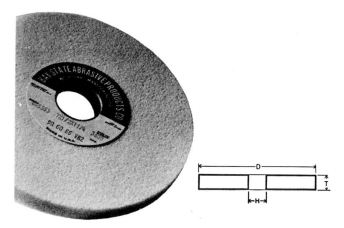

Straight wheel for grinding on periphery. Has a straight face, side and hole.

Fig. 3–15. Commonly used grinding wheel shape types. Types 1, 5 and 7 are for peripheral grinding; types 2, 6, 11 and 27 are for side or rim grinding (See also pp. 78, 79, 80.) (*Bay State Abrasives, Dresser Industries.*)

TYPE 5 — RECESSED ONE SIDE

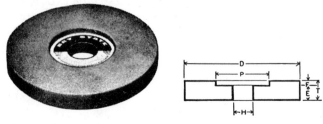

Recessed wheel (one side) specified DxTxH— Rec. 1/s DxF (diameter of recess x depth or recess). This is similar to a Type 1 wheel for peripheral grinding and allows for a thicker wheel to be used providing clearance for the flange and nut.

TYPE 7 — RECESSED TWO SIDES

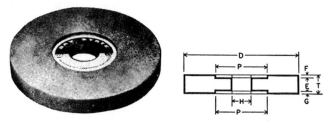

Recessed wheel (two sides) similar to Type 5 shape, but with two recesses of equal diameter and depth of recess is either equal or not as required for mounting extra thick wheel.

TYPE 2 — CYLINDER

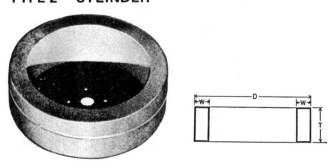

Cylinder wheel where I.D. (hole) is nearly as large as the O.D. Grinding is performed on the rim or wall end of the wheel.

Fig. 3-15 (*continued*)

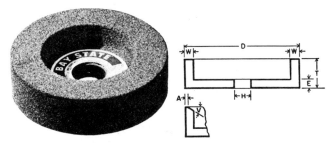

Straight cup wheel; similar to Type 5 except recess is much deeper. Grinding is performed on the rim or wall end of the wheel rather than on the periphery.

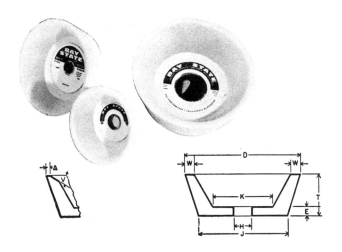

Flaring cup wheel is similar to Type 6 but with a tapered outside diameter to provide clearance of the work piece on certain types of grinding applications. Grinding is performed on the rim of the wheel.

Fig. 3-15 (*continued*)

TYPE 27 – RAISED HUB DISC WHEEL

Raised hub disc wheel (sometimes called "depressed-center" or "hat" wheel) is a reinforced organic bonded product used on off-hand, portable grinding applications.

Fig. 3-15   (*continued*)

backing plate without the danger of bolt heads interfering with the grinding action.

The basic shape of rim-grinding wheels is the type 2, or cylinder, wheel, also with three dimensions, except that because the hole is so big in relation to the diameter, the third dimension is the wall thickness rather than the hole. Wheel diameters range, in standard sizes, from 8 to 20 inches; in thickness, from 4 to 5 inches; and in wall dimensions, generally from 1 to 1 3/4 inch. Cylinder wheels are used almost exclusively for surface grinding.

The other major group of side-grinding wheels are the cup wheels—type 6, a straight cup; and type 11, a flaring cup. Portable grinding, for very rough jobs, is limited to organic bonds only. Cup wheels are also used in vitrified bonds for tool grinding. Portable grinding cups wheels are all 6 inches or less in diameter. However, wall thickness for portable grinding wheels is much greater (3/4 to 1-1/2 inch) that it is for tool grinding wheels (type 6, 3/8 inch; type 11, 1/4 inch).

Another side-grinding wheel warrants brief mention. Type 28 are rather thin depressed-center (or raised-hub) wheels, 7 or 9 inches in diameter, and

Fig. 3-16. Grinding with a type 27 depressed center wheel. Holding the grinder at an angle limits the grinding face to about 2 inches. (*Bay State Abrasives, Dresser Industries.*)

made only in reinforced resinoid (BF). They are used on portable grinders for weld grinding and other rough jobs. The offsetting of the hole permits these wheels to be conveniently bolted to backing plates. There is a similar wheel, type 27A, that is used for cutting-off.

The effect of the type of grinding wheel used on the ground surface should be mentioned. The scratches made in the piece-part surface by a peripheral grinding wheel are essentially straight and parallel. But the surface produced by any side-grinding wheel is essentially a pattern of overlapping arcs, which is sometimes known as a "dutch" surface. This is frequently considered a desirable appearance for the finished product.

Standard Shapes of Grinding Wheel Faces

Although we customarily think of a peripheral grinding wheel as having its grinding face at right angles to the side, grinding-wheel suppliers are prepared to furnish other "standard" face shapes, as designated by the American National Standards Institute (ANSI B74.2–1974) or the International Standards Organization (ISO), as shown in Figure 3–17. If a plant uses any quantity of formed wheels with any of these standard faces, it may be more economical to order the shaped wheels rather than to order wheels and shape them in the plant. Wheel faces are likely to be a little more consistent if ordered from the wheel manufacturer.

Mounted Wheels, Points, Cones, or Plugs

Another group of wheel shapes is one that includes abrasive wheels or points which have mandrels cemented or molded in so that the wheels can be held in a chuck as if they were some kind of drill; or cones and plugs, which have a blind-hole threaded bushing for mounting. Cones and plugs

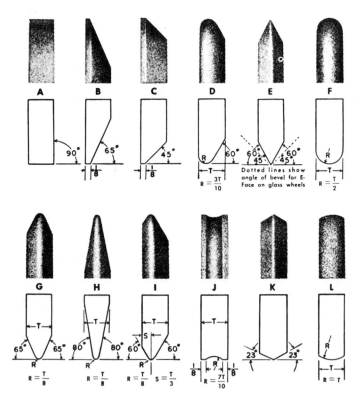

Fig. 3–17.   Wheel face shapes for straight wheels, available from manufacturers. (*Bay State Abrasives, Dresser Industries.*)

are used in portable grinders for rough foundry snagging of castings; mounted wheels and points are used mostly for touch-up work, or for internal grinding. Since their diameter is at most 3 inches, and often only a fraction of an inch, the wheels must be rotated at high revolutions per minute (rpm) to achieve an efficient surface speed in surface feet per minute (sfpm). The rate of these grinders (Fig. 3–18) sometimes exceeds 100,000 rpm.

Cutoff wheels, sometimes called abrasive saws, should probably be considered separately as a group, even though they are technically very thin resinoid or, in small, very thin sizes, rubber-bonded type 1 straight wheels. They function as abrasive saws, rather than as grinding wheels, for parting materials that are too hard to cut with bandsaws or similar sawing methods. (Fig. 3–19) In nonreinforced bonds, they range in thickness from 1/16 inch and 6 inches diameter to 1/4 inch and 34 inches in diameter. The range for reinforced bonds is from 6 inches × 1/8 inch to 48 inches × 3/8 inch.

Abrasive cutoff or sawing is the only material-parting process that produces two finished ends with virtually no burring to be removed later. It is thus particularly desirable for high-production cutting, mainly on harder materials, since automatic raising and lowering of the wheel, infeed of the

Fig. 3–18.   Grinding with a mounted wheel, also with a portable grinder. (*Hitchcock Publishing Company.*)

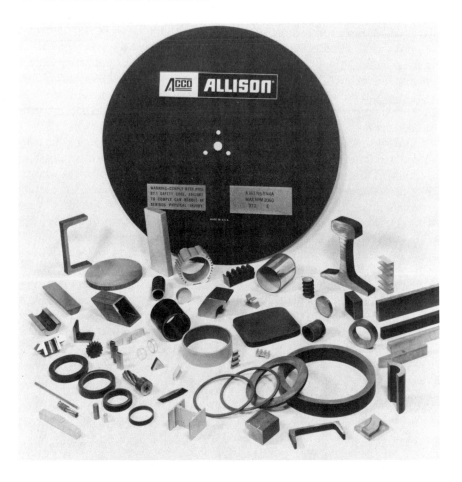

Fig. 3-19. An assortment of parts cut off with a thin abrasive wheel. Abrasive cutting like this leaves no burrs. (*ACCO Industries, Inc.*)

rod or wire from which the parts are cut, and ejection of the finished parts are simple machine-design problems.

Sticks, Stones, and Segments

Mention was made earlier, in connection with vitrified bonds, that abrasive sticks and stones, like the pocket sharpening stones that are common souvenirs at trade shows, can be "shaved" from pressed blocks of green or unfired vitrified mix. This is the standard method for making such abrasive tools, because it is much faster and produces a better-quality product than would probably result from individual molding. But, segments (Fig. 3–20) used in specially designed holders (called chucks) for vertical-spindle surface grinding are a somewhat different story. Because of their odd shapes, standard pressing methods can not be used. For one thing, the inventory of molds required would be excessive. For another, and more important, factor, the shapes would make it difficult to produce a segment with equal grade throughout; the segment would tend to be hard on the outside and softer on the inside. However, because of the large area of abrasive-work contact, only soft grades are needed—mostly D through G or H at the outside, with *very* open structure—11 through 15—so that high pressure is not needed. Grinding with segments will be considered in more detail in the chapter on surface grinding. Segments are very efficient and effective abrasive tools.

Diamond and CBN Wheel Shapes

Diamond and CBN shape symbols bear some resemblance to those for conventional grinding wheels, but they look more complicated because they tell more about the wheel. Most of the similarities stem from the fact that the diamond wheels were developed after the other symbols were well established, so it was logical that both should have type 1 straight wheels, type 6 straight cups, and type 11 flaring cups. For diamonds and CBN, there are 9 basic core shapes that can be made in 35 different variations, many of them of the type 1 shape.

Fig. 3–20. A common type of segment. The pads on top help to cushion the segment in its holder. (*Bay State Abrasives, Dresser Industries.*)

GRINDING ACTION

Grinding action is a term used in the abrasive industry to express the overall efficiency of a grinding wheel. Many factors can affect a wheel's performance, not all of which are related to the wheel itself. Some, in fact, may be quite remote, particularly those that cause vibration, which is the arch foe of good grinding. One problem is that frequently there is no clear knowledge of how a change in one factor affects the others. What is known, of course, is that some seemingly small changes in wheel specifications or in other grinding conditions can have a disproportionate effect on a wheel's total performance—its grinding action.

Wheels

A continuing complaint of suppliers' representatives and engineers is that when there is trouble, the wheel gets an unfair amount of the blame. Let a grinding operation go a little bit sour, and everyone, say the representatives and engineers, begins to fiddle with the wheel specification, when more attention should be paid to the grinding machine and its condition, to the coolant, and perhaps even to the machine's location.

Of course, the wheel is a definite object. It is made with a known kind of abrasive grain within a given size range. Its bond is known, though its particular variety of, say, vitrified bond may not be common knowledge. Its grade and porosity are also known quantities. The specification is a starting point. (Fig. 3–21)

But if a wheel of a given specification has been performing adequately for a time before the trouble arises, it is worth while to look elsewhere when trouble comes.

(For example, there is a consensus among grinding-wheel experts that many grinding machines in this country are underpowered and not sufficiently vibration-free to provide topnotch grinding performance, and that machine parts, particularly wheel spindles, are used past the point of top performance. It is a truism of the field that vibration of any kind demands a harder wheel, which is usually less desirable overall. But this is not a discussion of trouble-shooting.)

Still another factor in grinding action is the speed of the wheel, together with the speed at which the piece-part travels. It is logical that the higher the speed of the wheel (within safe limits, of course), the greater the number of abrasive "teeth" that pass across the surface to be ground, and the greater the amount of stock removed. Some research into speeds above those currently considered safe has shown that there are plateau ranges, where increases in speed do not increase stock-removal rate, but eventually

Fig. 3-21. The color of a wheel often gives an indication of its formulation, as in these white wheels. But the standard marking on the wheel blotter is a better one. (*Bay State Abrasives, Dresser Industries.*)

further increases in wheel speed result in greater stock removal. Currently these are matters of interest in research rather than in production.

Coolant

Another gray area in the matter of grinding action is the influence of the grinding fluid or coolant. There are many instances, reasonably authenticated, in which a change of coolant produced a significant improvement in the grinding action of wheels on a particular application. Naturally, no one ever publicizes the changes that go in the opposite direction. And the decision about whether or not it is better for a plant to use a central system with the same coolant for all grinding operations (which may mean that the coolant is something of a compromise but that removal of heat generated in grinding is easier, for one factor) is not one to be made lightly. Using only one coolant also simplifies storage, probably reduces the per-gallon price, and so on; but obviously the coolant selected can not be the best one possible for all the operations. This topic will be discussed at greater length in the chapter on grinding fluids; all that is intended here is to point out that the coolant used, along with the machine and its condition, the general conditions surrounding the machine, or indeed, the work material itself, all have an effect on the efficiency of the operation, and in the event that efficiency declines, one must look at any recent change in any of the factors just listed to try to determine what effect it may have had. For one example, if piece-parts come to a grinding operation with more stock to be removed than had been the case before, then the grinding operation is going to look bad unless this change is compensated for.

Wheel Selection

Selection of a grinding wheel of a particular specification for a particular application can be tricky, because one can not as yet read the correct specification from a list. Of course, as was pointed out in the discussion about the possible number of combinations of the various factors, the first gross elimination of possible specifications is relatively simple. The broad abrasive type is usually pretty well indicated; grain size can be narrowed down to perhaps three or four adjacent sizes. Grade will probably be similarly limited. Structure is standard in many specifications, based on grain size and grade. And gross selection of a bond is easy, but when the choices narrow down to two or three similar vitrified bonds, for one example, then the selection task gets rougher.

The actual selection may be made by a supervisor or engineer who is close to the job, but it is often left to the expertise of a distributor or company salesman. Grinding-wheel company salesmen, particularly the experienced ones, see a great many jobs and a great many machines, and most of them try to give their customers the best wheel that they have for a particular application.

However, if there is a prospect of a larger order, or of continuing business, most wheel companies will supply, say, half a dozen "trial" wheels which can be used on the job to compare with other wheels. This used to be fairly common practice; but with increased costs and greater performance predictability, the requirements for trials may have been tightened up, though it costs nothing to ask for free trial wheels.

If your company does get wheels for a trial from two or three suppliers, it is in your best interest to make the test as even-handed as possible. (Obviously, if any one is predisposed to the wheel supplied by one company, it's less than fair to the others to have them submit wheels for a contest that has already been decided.) Here are some suggestions for proper testing procedures:

1. The scope of the test must be in keeping with wheel usage. People tend

to think of testing as involving a lot of paperwork, records, and the like, but this need not be the case. It may be as simple a thing as keeping track of the number of parts ground per wheel, or ground between wheel dressings (resharpenings). It could also involve some detailed accounting, not only of wheel wear, parts per dress, and so on but also records of the surface(s) produced, the burn, or some other undesirable condition. If the wheel is in intermittent use only, all that may be needed is a record of when a wheel was mounted and when it was discarded. Perhaps the key is to do the simplest test that will give a satisfactory ranking of the test wheels.

2. Grinding conditions must be kept identical, so far as it is possible to do so. This usually means that the tests will be carried out on one machine, using the same coolant for all wheels, and preferably, if conditions warrant, using the same operator. In other words, for accurate wheel testing, only the wheels should be changed. And incidentally, to avoid unconscious—or sometimes conscious—operator preferences, it may be desirable to identify the wheels only by an alphabetical code or some other similar device. A skilled operator can skew test results if he is so inclined.

3. Test conditions must be as near actual operating conditions as is possible to ensure that the test results will carry over into actual production.

4. The basis of the test must be clearly delineated, for this is most important. Some, such as the time that the wheel remains useful, or wheel life, have been mentioned earlier. Production per wheel or per dress is a common basis for testing. Any reduction in dressing time, which is nonproductive time, is also a desirable factor to consider. For example, if one wheel will grind 20 pieces before it needs dressing and another wheel will grind 25, the second wheel has a substantial edge.

In fact, if your criterion is the number of good parts that the wheel produces before it reaches stub or discard size, then the dressing procedure must be kept to a standard. Wheel dressing is essentially induced wear to uncover new and sharp grain, so it does not take a very much longer time per dress to increase wheel wear substantially.

Wheel Dressing and Truing

Dressing and truing may easily be the two most confused terms in the grinding wheel business, and both affect performance. They are done with the same tools, and in the same manner. Only their objectives are different. *Dressing* is an operation done to restore or sharpen a grinding wheel face (or side); any shaping or squaring or restoring of the concentricity of the wheel as it is mounted is a bonus. *Truing* involves establishing or restoring concentricity of the mounted wheel, or shaping or forming its grinding face; any sharpening of the wheel face during the process is accidental. But a wheel that has been trued (i.e., made concentric with the machine spindle) is likely to be dressed; the reverse is less likely to happen. The confusion is not lessened by the fact that any forming of the wheel face is commonly called "form dressing" rather than "form truing."

For another distinction, truing is done only when a wheel is mounted (or remounted) on its machine spindle or other mounting device, because the wheel must be concentric (true) with the machine mount, not with itself. If the wheel-holding device is rigidly held to the wheel, then the wheel may be removed and replaced on the machine as many times as need be without re-truing. But if the wheel is freed from its holder, as happens when it is removed from a machine spindle, then it must be trued again with it is replaced.

Ideally, it would seem desirable to use a wheel so closely adjusted to the application that the force of grinding would be enough to pull dull grains out of the wheel face, but such a close specification is difficult to find. Most likely, the wheel would wear faster than is economical or, if the job involved a long pass across the piece-part surface, faster than is desirable for flatness. In this kind of flat grinding, if the wheel wears too fast, the finish end of the pass will be higher than the beginning end, and the piece-part will not be flat. On the other hand, a wheel that is too hard, in terms of grade, will not wear; it will hold the grains after they have become dull and stopped cutting. Longer wheel life is not desirable either. The usual compromise is to get a wheel that is a little harder than the ideal, and dress it as needed to maintain its cutting ability.

Before discussing the tools for dressing, a quick consideration of the conditions that the operation corrects is appropriate. One is the condition mentioned earlier, where the grains in the wheel face are retained after they become dull and are no longer cutting. In such a condition, the grains have the additional negative action of rubbing and heating up the piece-parts, which is also undesirable. If the operation is done with a grinding fluid or coolant, heating is reduced, but not eliminated. And because heat is inherent in a grinding operation by its nature, additional heat, with its negative effects, is not needed. The smooth wheel face is called *glazed,* and can be corrected only by dressing the dulled grains from the wheel face.

The other wheel-face condition that hinders cutting is *loading,* which occurs when bits of the work material become lodged in the wheel face. (Fig. 3–22). This does not happen when the wheel face is sharp, and it is retarded when there is sufficient coolant under enough pressure to remove chips from the wheel face. As grains become dull, loading becomes more likely, and so does glazing, but the two conditions do not necessarily go together.

Fig. 3–22. The dark spots on the wheel face are loading from bits of work material picked up by the wheel. Glazing shows up as a smooth, shiny surface on the wheel face. (*Bay State Abrasives, Dresser Industries.*)

If loading occurs when the wheel is fresh and sharp, a poor selection of wheel has been made. It is preferable to change to a more open wheel without delay.

There is one circumstance where dressing is done intentionally to dull the grain. As has been noted, sharp abrasive grain cuts and dull grain polishes, thereby producing a better finish. So in cases where it is more economical to rough and finish the piece-parts with the same wheel, this can be done by intentionally dulling the grain in the wheel face with the dressing tool after the rough cuts have been made. With skillful dressing it is possible, according to at least one expert, for a coarse-grain wheel—say 36 or 46 grit—to produce a finish like that customarily associated with a 240 grit size wheel.

The principal concern of management, however, is frequency of wheel dressing rather than methods. It's an old truism of abrasive use that more abrasive is dressed away than is ground away. Like most old truisms, it has more than a grain of truth. But it must be kept in mind that dressing time is not productive time (except when dressing is automatic and done while the wheel is grinding), that abrasive grain dressed away is lost, and that excessive dressing is a waste of time and abrasive.

Tools and Equipment for Dressing

The tools are the same whether the operation is truing or dressing, and so are the procedures. It is the purpose that differs. Dressing, as was stated earlier, is primarily a sharpening of the wheel's grinding face. Truing involves the additional aim of ensuring that the grinding face is concentric with the center of the spindle on which the wheel is mounted. Truing is most important in the operation of diamond and CBN wheels, where anything beyond a minimum of dressing is extremely expensive.

Tools for dressing fall into four categories: mechanical, diamond, crushing, and abrasive. Diamond dressers are the most widely used, and come in a variety of styles.

Mechanical Dressers. These are most frequently spurlike metal "stars" which are mounted loosely on a pin in a holder, though there are other designs. When the dresser is held against a rotating wheel, the spurs rotate and pick out of the wheel face both dulled grains and bits of work material. These dressers are used primarily on rough grinding wheels. (Fig. 3–23.)

Diamond Dressers. Diamond dressers, which are by all odds the most numerous type, range from the traditional hand-held type to mounted dressers (Fig. 3–24) to the form dresser (Fig. 3–25).

The original diamond dressers were mostly large stones which were not, obviously, of gem quality. The high cost of stones and the size of the loss if one came loose or were otherwise misplaced, has caused a major switch to a cluster type of smaller diamonds imbedded in a matrix. The clusters probably do as good a job as the large diamonds, and are certainly less vulnerable to loss. Diamonds are also mounted on simple swing-type fixtures to dress either a convex or concave radius on the periphery of a wheel. On reciprocating-table surface grinders, a diamond is often mounted on the

Fig. 3-23.   Mechanical dresser in use. Such a dresser is for use on coarse-grit wheels for rough grinding. (*Desmond-Stephan Mfg. Co.*)

Fig. 3-24.   Closeup of a diamond dresser, of which there are many styles, because diamonds are used for almost all precision dressing. The diamond (or sometimes a cluster of diamonds) is in the point, up against the wheel face. (*Desmond Stephan Mfg. Co.*)

magnetic chuck. And the form dresser shown (Fig. 3–25) could also be so mounted.

The diamond dresser is usually traversed back and forth across the face of the wheel. If the traverse is slow, the diamond tends to cut the grain and leave it partially dulled, producing a wheel face more adapted to finishing than to cutting. A faster traverse produces a sharper, faster-cutting wheel. Ability to dress the wheel properly is one of the marks of a skilled grinding-machine operator.

Crush Dressers. Crush dressing is a technique which involves pushing the grinding wheel into the formed dresser with such force that the face of the wheel actually takes on the shape of the dresser. (Fig. 3–26.) This means that the machine must be rigid enough to withstand substantial force, a factor that eliminates many of the lighter-built machines from consideration.

Many installations, for instance, horizontal-spindle, reciprocating-table surface grinders, have two crush rolls. One is mounted on one end of the table as the work roll, which dresses the wheel during production. The other, called the "master roll," is on the other end of the table. When the work roll becomes worn, as it does eventually, the operator simply moves the wheel to the master roll for redressing and then regrinds the work roll to shape.

Crushing produces a wheel face which is very sharp and free-cutting. Such a face enables the wheel to cut parts from a casting or other blank without prior machining. In fact, this one-step machining to a finished part is an example of the many parts which can be produced in one setup

Fig. 3–25. Diamond dressing block to form a bonded abrasive wheel by cutting the grains. Such a block can be mounted on the magnetic chunk of a surface grinder. (*Engis Corporation.*)

Fig. 3–26. Crush dresser mounted behind the wheel. For dressing, the wheel is fed back into the roll with considerable pressure, enough to break down and sharpen the grains. (*Bendix Corp.*)

without any intervening transportation or other costs. Labor time is usually lower too.

Abrasive Wheel Dressers. Some abrasive wheels, usually vitrified silicon carbide, are sometimes used on cylindrical and centerless grinding wheels which are large enough that a diamond might leave dressing marks that would spoil the part surface. An abrasive wheel as a dressing tool is effective, leaves a sharp wheel face, and requires only a little skill. Abrasive wheel dressers are also used on diamond wheels—a case of the inexpensive softer wheel cutting the expensive harder wheel—but there is usually only a little diamond to be removed anyway. Considering the price of diamond, it is usually more advantageous to have the operator take some extra time to mount the wheel true in the first place, rather than to remove abrasive later to true the wheel, which is a necessity with both silicon carbide and aluminum oxide wheels.

Wheel Surface Speed

"Surface feet" and the abbreviations "sfpm" or "fpm" are the abrasive wheel industry's methods of identifying a very important concept: the speed at which a point on the surface of a grinding wheel is traveling. It can be considered comparable to miles per hour in rating automobiles. Its importance stems from two limitations: first, any grinding wheel has to attain some minimum speed in order to cut efficiently; and second, exceeding the established safe speed limit is a safety hazard for the operator, and very likely for the people in the area around the grinding machine. The safe, efficient operating rate for any grinding wheel is between these limits.

An important consideration is that most grinding machines are designed with a fixed spindle speed (rpm) in mind, although some can be furnished with a variable speed spindle. Usually the spindle speed and the size of the grinding wheel are factored in so that the wheel's surface speed is in bounds. However, with wheel wear, there is a reduction in wheel speed, which is a factor in determining the diameter at which the wheel should be discarded. The basic formula is

$$\frac{\text{Diameter (inches)} \times 3.1416 \times \text{rpm}}{12} = \text{sfpm}$$

Thus, a 20-inch wheel running at 1240 rpm would be traveling at about 6500 sfpm, the generally accepted safe speed for a vitrified wheel. But if the wheel wears to 16 inches in diameter, and the spindle speed remains at 1240, the wheel is traveling at only about 5200 sfpm, which is safe enough but probably inefficient.

SUMMARY

Bonded abrasive wheels and other shapes together constitute an efficient, adaptable, and often less-expensive means of machining a range of materials from most of the soft metals to most of the hardest materials known. Furthermore, these wheels will remove stock effectively at a much faster rate than they are usually given credit for. For some shops, grinding is used only when there is, say, about 1/8 inch of stock to be removed; with more stock, the tendency is to mill or turn the parts. However, others routinely grind stock up to 1/2 inch.

It is somewhat odd that grinding wheels, which are the preferred stock-removal tools in foundries and steel mills where stock must be removed efficiently and cheaply from castings and billets or blooms, and which are also the preferred if not the only method for producing very close-tolerance dimensions and super-quality finishes, have not made more headway in the middle ground where they compete with cutting tools. But that is the case, and one of the objectives of this book is to attempt to change this.

4

Coated Abrasive Products

In the coated abrasives business, two of the three major suppliers are also principal suppliers of abrasive grain, whereas the third is much more noted for its expertise in applying adhesives to backings. An interesting highlight of the fact that coated abrasive products have three principal components—abrasive and a backing material held together by an adhesive.

The idea of gluing abrasive grain to a backing is not new, as the now-outdated but still frequently used terms "sandpaper" and "emery cloth" suggest—outdated terms because there is no sand used today for coated abrasives and only a little emery. The first combinations of these three basic components—abrasive, backing, and adhesive—may very well have taken place before 1800. The first known magazine article about coated abrasives dates from about 1808, at least three decades before the patenting of the first grinding wheel in 1842.

The development of coated abrasives for the machining of metals was delayed for years because of the lack of waterproof adhesives (which meant that belts and other shapes could not be used with coolants) and the lack of stability and rigidity in the machines. Coated abrasives on light-duty machines became established for woodworking. But their use in the machining of metals spread rapidly with the introduction of waterproof adhesives and the design and manufacture of more-massive and more sturdily built machines with power on a par with the power used for abrasive wheels.

So today coated abrasives, mostly belts, may be regarded as capable of a wide range of processing from cleaning eggs to machining metals such as cast iron and aluminum; as well as some much harder, more rugged jobs such as steel-mill billet cleaning, descaling, and conditioning. The grain on a belt cuts material. (Fig. 4-1.) This discussion will be much more concerned with the middle part of the range, where abrasive belts are providing increased competition with both cutting tools and grinding wheels.

BELTS, CUTTING TOOLS, AND WHEELS

A coated abrasive belt is a length of coated abrasive material spliced together on the bias to make as smooth a joint as possible. (Fig. 4-2.) It is mounted over two and sometimes more pulleys—one called the contact roll or wheel, which forces the belt against the work; and the other an idler pulley, or sometimes the power pulley, depending on the design of the machine. (Fig. 4-3.) The contact roll can also be the driving roll. When the belt is used to machine parts, it makes chips such as those illustrated (though one-twentieth the size shown or smaller). It is their chip-making ability, which they share with grinding wheels, that makes belts competitive in machining.

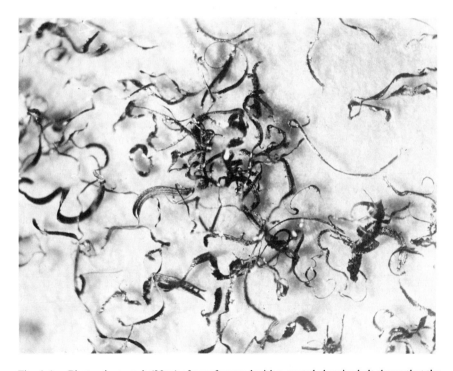

Fig. 4-1. Photomicrograph (20×) of swarf ground with a coated abrasive belt shows that the individual grains cut like turning tools or milling cutters. The precise orientation of the grain on the backing improves the belt's cutting ability. (*Norton Co.*)

Fig. 4-2. Spliced belt, with arrows showing the two ends of the splice. Joint is virtually smooth, so that it will leave no marks on the finished surface. (*3M Company.*)

Belts vs. Cutting Tools

The theoretical case concerning abrasive belts and cutting tools can be stated very quickly. On the surface of a 4- × 96-inch long 50-grit belt there are about 500,000 abrasive grains. At average belt speed, each of these will come in contact with the work more than 600 times per minute. At each contact, each grain theoretically removes a minute chip of the work material. This makes a potential stock-removal capability of some 300 million chips per minute. Obviously not all the grains remove chips; but if it is assumed that only a million tiny chips are removed per minute, while a much larger single continuous chip is removed by a lathe tool, then it is possible that the million tiny chips outweigh the one large chip. A similar analogy applies to milling cutters or planing and shaping tools.

Belts vs. Abrasive Wheels

The principal advantage of belts over wheels is the belts' ability to grind in one pass, widths up to 48 inches without lengthwise splicing—and wider widths with splicing. (Fig. 4-4.) Moreover, it is very easy to set up several belt grinding heads on the same machine, so that the piece-parts are

Fig. 4-3. This is the simple contact wheel-idler pulley setup which is the basis of all coated abrasive belt machine design. The contact wheel, at the right, is usually larger and is serrated. Some machines may have two or more idler pulleys. (*Hitchcock Publishing Co.*)

roughed and finished in one pass as they go under the heads on a conveyor belt. Belts, however, can remove significant stock. (Fig. 4-5.)

Belts are rarely dressed; for the most part they are used until worn and are then discarded. On a multihead machine as described above, a belt might be moved from the first head to the last before being discarded. So the comparison is between dressing time and belt-changing time, and such a comparison often favors belts.

On the other hand, belts are pretty much limited to straight-line rather than form grinding, although some contouring is possible with formed contact rolls. And it is generally held—even though abrasive belt proponents would argue the question—that belts cannot achieve quite the precision that is possible with wheels. There is some "give" in the backing material and some in the contact roll, which makes the contact more spongy than that of a vitrified grind wheel and probably a little less precise. However, belts routinely can attain closer tolerances than can metal cutting tools. And if the contact roll is made of steel rather than the usual hard rubber, the precision of the operation is enhanced. Finally, though belts do not have to be dressed as a general rule, they do have to be changed more fre-

Fig. 4–4. Final polishing and oiling of stainless steel plate. Plate is 58 inches wide. (*Hitchcock Publishing Co.*)

quently than grinding wheels, because there is essentially a single layer of abrasive grain on any coated abrasive belt.

No Standard Marking System

A few of the coated abrasives manufacturers have worked up consistent marking systems of their own, though none can compare with the range of the standard marking system for grinding wheels. Nor is there an equal need for one. The total number of companies making grinding wheels, including those making diamond and CBN wheels, probably totals close to 100; the corresponding total for belts is approximately six. The com-

Fig. 4-5. The shower of sparks demonstrates the stock-removal ability of belts on cast iron. Removal rate is 1/2 cubic inch per minute. (*3M Company.*)

parative capital investment needed for the latter explains the difference, at least in part. The coated abrasive "making" machine is a large investment indeed. At any rate, there is plenty of space on the back of almost any piece of coated abrasive to print all the explanation that is needed, this reduces the need for a code.

Abrasive Types

For all coated abrasives virtually all the abrasive grain is either silicon carbide or aluminum oxide, augmented by a little natural garnet or emery for woodworking, and with minute amounts of diamond or CBN. Around 1935, about half the product was coated with natural abrasive and half with manufactured abrasive. Twenty-five years later, the manufactured abrasive share had risen to about 70 percent; and 20 years later, by 1978 or 1979, to over 90 percent. In fact, since most of the garnet- and emery-coated product goes into sheets for home workshops and do-it-yourself projects, industrial usage must be approaching 100 percent manufactured grain.

Aluminum oxide intended for coated abrasives is more elongated and pointed or sharp than is grain for wheels. As has been mentioned, the process for making coated abrasives permits orientation of the grain with its long axes at right angles to the backing, but the molding process for wheels does not include grain orientation. Aluminum oxide is a little softer but tougher (i.e., less likely to fracture under pressure) than is silicon carbide, and it is a more effective abrasive on all ferrous metals (except perhaps cast iron), on alloy steels, on tough bronze, and on hard woods. It is available in all grain sizes, generally from 600 (fine) to 12 (coarse). The size designations have the same general meaning for coated abrasives as they have for bonded.

Silicon carbide is regarded as a harder and sharper, though not tougher, abrasive than aluminum oxide. These qualities make it superior for use on low-tensile metals, glass, plastics, fibrous woods, leather, and other comparatively soft materials. It penetrates and cuts fast under light pressure, which is beneficial for polishing or other cosmetic surface treatments where appearance is more important than close tolerances. Silicon carbide is available in the same range of sizes as aluminum oxide, 600 through 12.

Backings

There are four general groups of backings used for coated abrasives: paper, cloth, vulcanized fibre, and combinations of these laminated together. Various weights of cloth are the most-used industrial backing.

Paper backings as a group are the least costly. They are generally used when strength or pliability of the backing is not critical—with sheets, for one example. The weight of the paper per ream (480 sheets of 24 × 36-inch paper) is indicated by a standard letter on the back of the sheet. The lightest, A weight, is 40 pounds per ream. C weight paper is 70 pounds; D weight is 90 or 100 pounds; and E weight, the heaviest, 130 pounds. Paper belts, which are used in some woodworking applications, are almost always E weight. A-weight paper may be called finishing paper, and C and D weight papers are sometimes termed cabinet papers.

Cloth backings are generally made from a specially woven type of cotton cloth. The two most common types are drills, marked with an X on the back, and jeans, marked with a J. Both have a twill weave. The principal difference is that drills are made from heavier threads, though there are fewer of them per square inch than there are for jeans, so that drills are the heavier and the stronger of the two. Drills are preferable for heavy work with coarse abrasives; jeans are better when more flexibility is needed.

The strength of the backing for belts, particularly, limits the pressure that coated abrasive belts will stand without tearing. And pressure, as has been mentioned earlier, is a key factor, along with speed, in the stock-removal capacity of any abrasive tool, whether wheel or belt. The consensus is that cloth belts have been pushed about to the limits that they can take; and that further increases in ability to withstand pressure will come from synthetic backings.

The problem is that heavier cloths, which are quite readily made, cannot also be made flexible enough to bend around the contact roll and the driver and possibly idler rolls that are the basic elements of any coated abrasive belt grinder.

Vulcanized fibre is a very heavy, hard, and strong backing, of limited flexibility, used principally for resin-bonded discs on heavy duty sanders, and in a thinner version for drum sanding. (In drum sanding, a woodworking application, a sleeve of coated abrasive is slipped over a round, expandible holder, so that the assembly resembles a single-layer abrasive wheel.) Belts are preferred in metalworking and, for that matter, in most woodworking, because the abrasive, as a result of the length of the belt, has more of an opportunity to cool off between passes across the piece-part(s). Vulcanized fibre is made by impregnating cotton rag base paper with zinc chloride and then vulcanizing five to seven sheets together by heat, before the backing is coated with adhesive and abrasive.

Combinations may be either paper and cloth laminated together or laminated fibre and cloth. Both backings are sturdy and shock-resistant, though not particularly flexible. The first combination is used mainly on high-speed drum sanders; the second is an alternative to fibre backings for sanding discs on portable sanders.

Adhesives

The obvious function of any adhesive is to attach the abrasive grain to the backing. In the manufacturing process for coated abrasives a continuous strip of backing 52 inches wide moves through each step of the way—backprinting, coating with adhesive (the "make" coat), coating with abrasive, a second coating (the "size" coat) of adhesive—until the strip is wound up as a "jumbo roll" about three feet in diameter and trimmed to 48 inches width.

If the product is intended for dry use, as, say, in most woodworking applications, then the adhesive can be glue—straight animal-hide glue, which is the traditional adhesive. With coolants, the adhesive can be one or another of the resins, which are basically liquid phenolics. For special uses, the two may be combined, either glue over resin or resin over glue. Resins, particularly, may be modified to provide shorter or longer drying times, greater strength, more flexibility, or other desirable properties.

METHOD OF MANUFACTURE

In the manufacture of coated abrasives as outlined above, the key step is probably the application of the abrasive, which may be done by pouring the grain in a controlled stream from above, or, as is more likely the case today, passing the strip of backing over a pan of abrasive with the adhesive-coated side down, and at the same time passing through the abrasive an electric current which causes the abrasive grains to project themselves upward along and parallel to the lines of force, so that they embed themselves point up in the adhesive-covered backing. This grain orientation produces a very sharp, fast-cutting abrasive tool.

The amount of abrasive grain deposited on the adhesive-covered sheet can be controlled to amazing accuracy by adjustment of the abrasive stream and manipulation of the speed of the sheet of backing. Of course, once these two factors are established for any one jumbo roll of backing, they are not modified during the coating of the roll.

There are two variations of coatings—closed and open. A closed coat is one in which the abrasive grain completely covers the coated side of the backing. This is generally preferred for severe service. An open coat (Fig. 4–6) is one in which individual grains are spaced out to cover from 50 to 70 percent of the surface. An open-coated abrasive is more flexible than a closed-coated product, and it is less likely to become clogged or loaded with bits of the work material.

From the making unit the product is carried by a festoon conveyor system through a drying chamber to the sizing unit, where the size coat of adhesive is applied. The two films of adhesive unite to anchor the grain securely. Following a second trip on a longer festoon conveyor through a drying and curing chamber with closely controlled temperature and humidity, the product is wound into jumbo rolls for storage and conversion to marketable forms of coated abrasives. (Fig. 4–7.)

As the coated abrasive sheet is unrolled from the jumbo roll it is quite stiff, because it is made up of two coats of adhesive applied over a strip of cloth, paper, or fibre. So the first step in converting it to marketable forms is to flex it, that is, to break the adhesive in a controlled process so that it won't break in a haphazard and uncontrolled fashion later. Flexing may take two forms, single and double. Single flexing involves break lines at 90° to the edge of the roll. With fine grits, the break lines are usually close

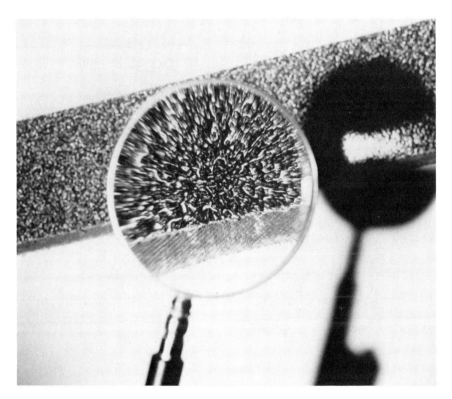

Fig. 4-6. Magnification emphasizes the backing showing through on this open-coated strip of coated abrasive. Coverage is 50 to 70 percent. (*3M Company.*)

together, thereby providing a softer flex. With coarse grits the break lines are farther apart, thus making a stiffer flex. This procedure leaves the sheet stiff in the crosswise direction, but with lengthwise flexibility that enables the sheet to conform to the arc of a drum or pulley without random, harmful cracking. The single flex is sometimes done at an angle other than 90°, but this is not common.

Double flexing involves breaking the sheet at two 45° angles to produce a criss-cross pattern of break lines, which makes the sheet flexible in almost any direction. Double flexes are required on coated abrasives intended for the sanding of shoulders and other contoured work. Triple flexing is a combination of the two; it makes the sheet extremely flexible for the sanding or grinding of very irregular contours.

Because any flexing breaks the continuity of the bond, it tends to decrease the durability of the product. For that reason, flexing should be held to a minimum (which for most industrial applications is a single flex)

Fig. 4–7.   Jumbo rolls of coated abrasive in storage. Each roll contains about 50 linear yards of product which is approximately 50 inches wide. (*3M Company.*)

so that the belt will conform to the arc of the contact roll or wheel and the other pulley(s) in the system. Minimum flexibility consistent with operating requirements is always the economical selection.

Slitting and Other Cutting Operations

Some coated abrasives are sold to consumers in the original 48-inch-wide, 50-yard-long jumbo rolls, but most are sold in narrower slit widths, as, for example, stock intended for belts. The slit rolls are produced on a machine that uses circular knives against a hardened steel roll and then rewound to the desired length. Most of the 50-yard rolls are continuous strips, though difficulties in manufacturing sometimes make it necessary to produce rolls made up of two or three lengths; but if this is the case, an extra length is included for each short piece, so that where a 50-yard continuous length is required, the roll can be spliced so smoothly that the joint is virtually invisible.

For continuous belts in any desired width and length, slit rolls of the proper width are cut to length, including an allowance for the joint overlap or splice. Most belts are spliced at a 45° angle, though narrow belts are spliced at a more acute angle, and wider belts at a greater angle. At the joint, to avoid a bump which would ruin the piece-part surface, the two angle-cut ends of the belt length are skived, that is, the layer of abrasive is removed from one end of the length and a minute layer of backing is removed from the other end, so that when the two overlapped ends are joined, the belt is of correct length and the joint is little, if at all, thicker than the belt. For coarse-grain belts where the prime objective is to remove stock, only one end may be stripped of abrasive (single skiving), because the surface will not be affected seriously by a poorer finish and because the whole strength of the belt is needed to withstand the high pressures in such operations.

Sheet and Disc Cutting

Standard sheet sizes (usually 9 inches × 11 inches) are cut on a ream cutter, which is a combination of a slitter and a flying knife. Discs are cut on a punch press with a die or dies. The operation is usually done on a complete length, as a jumbo roll or a slit roll is unwound ahead of the punch press. Each disc with its center hole and any required radial slits is punched out in one operation for standard diameters.

ELEMENTS OF BELT GRINDING HEADS

Basic Design

Coated abrasives are used in many forms: sheets for sanding; discs with stiff backs to be used on portable grinders for rough applications, such as weld grinding; discs adhered to backing plates for wood sanding; small strips wound up and glued together like mounted bonded wheels; and continuous-belt grinders on which the work is either pressed against the belt backed up by a contact wheel or a flat platen between the pulleys or simply held against the belt without any backing (slack-of-belt grinding). But the major industrial grinding-head design is one in which the belt passes around two or more pulleys, with one of them, known as the contact roll or wheel (depending on its width), forcing the belt against the piece-parts. Or, as with a floorstand or backstand grinder—two terms of the same machine—the piece-part is forced against the contact wheel by the operator.

This basic assembly of drive pulley and contact wheel can be augmented by one or more idler pulleys around which the belt can pass. The whole idea of belt grinding, as opposed to drum sanding where the sleeve of coated abrasive is fitted snugly around a head, is that the travel time of the belt between contacts with the work gives the belt time and opportunity to dissipate grinding heat. A drum sanding head provides no such opportunity. So it is not uncommon to have more than two pulleys for the belt, but two would be considered the norm. (Fig. 4–8.)

Contact Wheels or Rolls

It should be clear by this time that effective stock removal with abrasive belts requires some kind of backup for the belt, and that the backup is usually a contact wheel or roll, though it could be, as noted above, a flat platen. Usually the platen is located between pulleys mounted one above the other, and there is frequently a horizontal work rest to help support the piece-parts.

The contact wheel is generally a grooved solid rubber tire (Fig. 4–3) vulcanized to an aluminum wheel or a metal rim mounted on a hub. (There are also sewn-cloth contact wheels, but they are primarily used for polishing operations without close tolerances or significant stock removal. (Fig. 4–9.)

Rubber contact wheels have become popular because of improved methods of bonding the rubber to the metal and because of improved

Fig. 4–8.   Another view of the belt system, with the guard raised so that the contact wheel and backstand idler can be seen. In operation, of course, the guard is closed. (*Divine Brothers Co.*)

Fig. 4–9.   This operator is polishing pistol frames with a 4- × 148-inch belt running over a compressed canvas contact wheel. (*Norton Company.*)

methods of controlling the hardness or density of the rubber. Hardness (durometer) of the rubber, along with the serrations, determines the cutting rate of the belt—other factors being equal. For a given grain size, a hard wheel makes the belt remove more stock than a soft wheel; and the harder the wheel, the faster the cut and the coarser the finish. So changing to a harder wheel has approximately the same effect as changing to a coarser grain size. In any multiple-stage operation, where two or more grinding heads are used, the hardest contact wheel that will conform to the surface and also produce a finish that can be blended in by the subsequent grinding is the one to be used.

Rubber contact wheels are manufactured in densities from extremely hard (90A durometer) to extremely soft (20A durometer). Wheels in the range of from 70A to 90A durometer are primarily for stock removal. Medium wheels (40A to 60A durometer) are for the type of operation involving some stock removal but primarily a good commercial finish. Soft wheels (20A to 30A durometer) are used for fine polishing and limited contour grinding. There is a third factor, however, that may sometimes come into consideration—the speed of the belt, and consequently, of the contact wheel. Increasing this speed will cause the contact wheel to act harder than it does at normal speed. And since most coated abrasive grinders are designed to run at a constant speed, adjusting speed might well be regarded as a measure to be used only in emergencies. The normal optimum approach is to select a contact wheel in keeping with production requirements, and with the grinder running at its normal speed.

The serrations or cross slots in the contact wheel also affect its action, and the aggressiveness of belt action and the coarseness of the finish increase as the size of the angle increases from a minimum of 15° to a maximum of 90°. At the maximum, 90°, there will be a very aggressive abrasive action, and subsequent coarse finish, along with a noise that sounds like a siren, which would probably be unacceptable as a working condition. There may also be an unattractive scratch pattern on the work.

Serrations produce alternating lands and grooves across the face of the contact wheel, and thus the ratio of the width of the grooves to the width of the lands, the depth of the grooves, the shape of the lands, and as mentioned, the hardness of the rubber in the contact wheel are all factors in determining the cutting action of the belt. The serrations also provide an easy means of controlling the breakdown of the abrasive grains to ensure renewed sharp grain edges. Beyond that, the flexing action of the belt as it passes over the lands and the grooves provides chip clearance and prevents the chips from loading the belt.

Belt Tension, Belt Speed, and Power Requirements

The tension on the belt, its speed, and the power requirements for abrasive belt grinding are all items of importance, because they affect the efficiency of the method and, in some measure, part of its costs.

Actual tension on a belt can range from a low of 4 to 5 pounds per inch of belt width (''inch of belt width'' is one of the common denominators of belt comparisons) to a high of 35 to 40 pounds. The lower range is used for contour polishing and similar applications, with the belt at reduced speed. In such circumstances it is preferable to use just enough tension to keep the belt from ''walking'' sideways on the contact wheel when the work is applied. It is also preferable for the belt to have 100° or more of uninterrupted wrap around the wheel before it comes in contact with the piece-part.

With harder contact wheels and heavier pressure on the piece-part, higher belt tension is needed. Unless steel contact wheels are used, high pressure deforms the contact wheel at the contact point, a condition which may cause the belt to pucker up ahead of the contact area if the pressure is high enough. Particularly if belt tension is insufficient, this puckering can cause premature "shelling," or stripping, of the abrasive, which in turn makes it necessary to discard the belt before its time. On the other hand, excessive belt tension can reduce some of the beneficial effects of the contact wheel.

Belts do operate at various speeds, of course, and like grinding wheels they have safe maximum limits, as indicated by the manufacturers. It is worth noting, though, that they have proved safe at a maximum practical speed of 10,000 sfpm, which is approximately 50 percent above the general safe maximum for vitrified bonded wheels, and about on a par for the general safe maximum for unreinforced resinoid wheels. Much of what is true about the effects of wheel speed and pressure on the work for wheel grinding is also true for belt grinding. And pressure of the work on the belt has the further beneficial effect of promoting breakdown of the abrasive grain and keeping the belt sharp.

Power requirements per inch of belt width range from 1/2 to 7 horsepower, with the lower end for very light work such as deburring, and the upper limit for operations requiring heavy stock removal. Probably most operations fall into the range between 1 and 3 horsepower, although the recent trend has been toward higher horsepower, because most experts agree that the majority of belt grinders, just like most hard wheels grinders, are underpowered for the most-efficient operation. As belt speed is increased, horsepower should be increased proportionately. The benefits of high-speed operation cannot be realized unless the horsepower available in the machine will maintain speed under load.

CONTACT WHEELS

Storage

Poor storage practices can do more damage to contact wheels than almost anything that can happen to them while they are in use. For example, on any operation involving the use of two or more contact wheels, it is usual to lean the extra wheels against the machine, a practice which can put the wheel into a permanent and undesirable set: it can produce both an out-of-round and out-of-balance condition that can mark the work, tire the operator, and shorten belt life. The best practice is to hang the out-of-use wheels on pegs through their center holes at locations convenient to the machine(s). Failing that, they should be stored flat on their sides.

Contact Wheel Selection

The selection of a contact wheel for a given application is obviously a job for someone with technical knowledge and a familiarity with the job, but the reader should have some understanding of the generalities on which the selection is made. Following are a series of statements that hold true for most of the grinding applications that use coated abrasive belts and contact wheels. Some of these have been indicated before; one or two of the ideas are new. Belt grit size is presumed to be a constant.

1. The harder the wheel, the faster the cut and the coarser the finish.
2. The smaller the wheel diameter, the faster the cut and the coarser the finish.
3. Finish and rate of cut go together. Coarse finishes accompany a high rate of cut; fine finishes, a lower rate of cut.
4. High speeds usually produce better finishes, except with soft wheels, for which speeds of 5,500 sfpm harden the wheel and make the finish coarser.
5. Higher groove-land ratios on serrated wheels increase the rate of cut and produce a coarser finish.
6. Serrated wheels reduce belt glazing and loading and give longer belt life than do smooth-faced wheels.
7. Hard contact wheels increase the stock-removal capacity of a belt, other factors being equal.

As Fig. 4–10 and 4–11 illustrate, the basic contact wheel-and-idler pulley (also sometimes called a backstand) is used on many different types of machines. The combinations can sometimes get rather complicated, but they can be used practically anywhere that a peripheral grinding wheel can be used, as is illustrated with the floor backstand grinder, the surface grinders, or the centerless grinders. In a centerless setup it is not uncommon to team up a belt abrasive grinding head with a rubber-bonded regulating wheel. And there are numerous such combinations of wheels and belts, some developed in-house and some developed by machine builders, to take care of special situations.

But while belts are probably the most-used coated abrasives, they are not the only ones. Stiff, heavy-coated abrasive discs (Fig. 4–12) compete with various kinds of grinding wheels on portable grinders. But one of the more

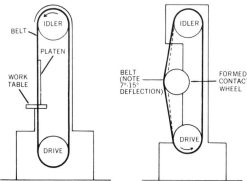

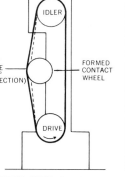

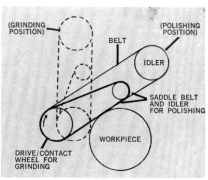

STATIONARY PLATEN GRINDER — For use where a true, flat surface must be developed. Uses: roughing, blending, finishing, polishing, and precision sizing.

FORMED WHEEL GRINDER — For use on contoured parts. Belt forms into contact wheel which is shaped to mate with workpiece. Uses: blending, finishing, and polishing.

CYLINDER GRINDER AND POLISHER — For larger cylindrical rolls and drums. Uses: precision sizing, blending, finishing, polishing.

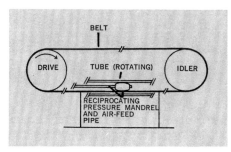

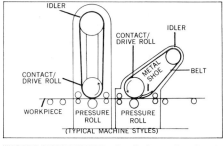

INSIDE DIAMETER (I.D.) TUBE GRINDER — For applying a finish to the inside diameter of tubes. Belt is spliced after being fed through tube. Work pressure is applied by means of a power-driven, air-inflated shoe which presses belt against inner wall of tube. Uses: finishing, polishing.

WIDE BELT GRINDER-POLISHER — For wide sheets, strips and panels. May be used over/under to surface both sides simultaneously. Uses: sizing, blending, finishing, polishing.

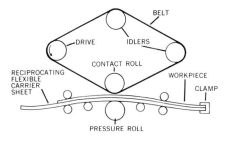

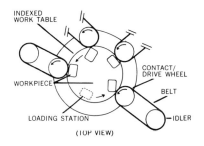

FLEXIBLE-BED SHEET GRINDER — Reciprocating carrier bed permits multiple-pass work for finishing and polishing on wide sheets of stainless steel where fine finishes are desired. Uses: finishing, polishing.

AUTOMATIC ROTARY-TYPE GRINDER — For production of identical workpieces — may use 6, 7 or more stations in sequence. Special tooling will accommodate most workpiece shapes. Uses: roughing, blending, finishing, polishing.

Fig. 4–10.   Variations in belt machine design. (*3M Company*)

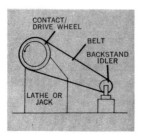

FLOOR BACKSTAND GRINDERS — For use on workpieces that can be carried and handled properly. This machine is basically a conversion of a floor lathe or polishing jack to abrasive belt use by means of a backstand idler pulley. Uses: roughing, blending, finishing, polishing.

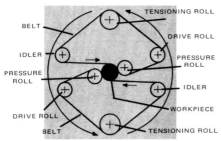

PLANETARY GRINDER & POLISHER — For use in grinding and polishing cylindrical, oval and some tapered stock in continuous or cut lengths. Note that the entire unit revolves around the workpiece as stock is traversed through the machine.

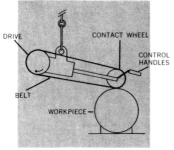

SWING GRINDERS — For use on larger workpieces that cannot be brought to a floor grinder. Available in 1-15 H.P. ratings. Uses: roughing, blending, finishing.

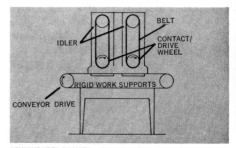

CONVEYORIZED GRINDER — For use on flat surfaces, including larger sheets and panels, where a succession of abrasive grades is used. Uses: simultaneous roughing, blending, finishing, polishing and precision sizing.

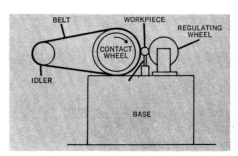

CENTERLESS (O.D.) GRINDER — For cylindrical shaped workpieces, such as rod or tubing of varying diameter and length, often requiring precision finishes. Uses: roughing, blending, finishing, polishing and precision sizing.

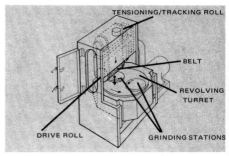

SURFACE GRINDER — For high volume stock removal operations. This machine features rotating fixture tables which are plunged vertically into an abrasive belt with extremely high pressure. This produces high stock removal rates and a flat surface. Note that the revolving turret creates a load-unload station while the alternate fixture table is in a grind cycle. Uses: heavy stock removal.

Fig. 4–11.   Variations in belt machine design. (*3M Company*)

116

Fig. 4-12. Portable grinding with a stiff coated abrasive disc cemented to a rubber backup pad. Note, however, that the pad does bend. (*Hitchcock Publishing Co.*)

useful recent developments is what is often termed a "flap wheel (Fig. 4-13.)" This is simply a number of partially slit sheets of coated abrasive clamped in a rotary holder, which can be mounted to revolve so that the abrasive strikes the work. The abrasive action depends on the grain size, the stiffness of the backing, and the rotating speed of the backing. Coarse-grained, stiff sheets rotating at high speed are effective in descaling or removing rust. Finer-grit and more flexible sheets at medium speed can be used for polishing contoured or formed parts. Sometimes slit sheets are used in a holder to polish grooves or slots.

WHEEL AND BELT GRINDER COMPARISONS

After having considered both grinding wheels and coated abrasives—principally belts—as forms of abrasive cutting tools, it is time to make some comparisons of the machines on which they are used, which are for the most part not too much different in design, although they may look considerably different to the layman because of the extra pulley(s) which are a requirement for any grinder using abrasive belts. It has been pointed out that grain orientation is much better on belts than it is on wheels, because

Fig. 4–13.   Coated abrasive "flap" wheel. Each set of three sheets has a sponge backup. With 100-grit abrasive, this wheel is probably for a semifinal finishing operation. (*3M Company.*)

the molding process for wheels does not at present make orientation possible. In terms of machine tool "rake," a metal cutting tool may be anything from positive to negative. The grain in a grinding wheel has only random rake, since the cutting angle of each grain depends on how it is positioned in the mix. But coated abrasive grain in general have various degrees of negative rake.

Peripheral or Side Grinding

Bonded abrasive wheels are designed generally for grinding on either the periphery or the side (rim). Of the grinding wheels definitely designed for side grinding or grinding on the flat, like type 2 cylinder wheels and types 6 and 11 cups, to mention the three most-used shapes, the area for grinding is usually small in relation to the outside diameter of the wheel. Segments in a chuck or holder for vertical-spindle surface grinding have greater area for flat grinding. And a bonded abrasive disc might be considered as a type 1 straight wheel without any center hole, although it may have bolt holes in a pattern on its flat sides for attaching it to a backing plate.

Belt grinding, at least all precision belt grinding, is peripheral grinding very similar to wheel peripheral grinding. But there is very little if any precision disc grinding with coated abrasives which is on a par with rim or side grinding with wheels. Belts have a somewhat higher surface speed than vitrified wheels—about 10,500 sfpm as compared with 6,500 sfpm for vitrified wheels or 9,500 for resinoid—and they have a much thinner layer of abrasive—essentially a single layer rather than the many in a conventional wheel. And belts must have the extra pulley(s), though that has been no barrier to rather frequent conversions of, say, cylindrical or centerless grinders from wheels to belts. The converted grinder may look a little odd, but it works.

Width of Grinding Surface

Comparative width of the grinding surface is another important consideration. Peripheral wheels for either surface grinders or cylindrical grinders range in width (thickness) up to about 6 inches for surface grinding and 4 inches for cylindrical. For centerless grinding, both the grinding wheel and the regulating wheel are wider, up to about 20 inches for standard wheels. Of course, on a centerless grinder, the duration of the grind depends on the thickness of the wheel, whereas both peripheral surface and cylindrical wheels can be traversed across the surface of the piece-part. Each passage of the wheel across the piece-part is known as a "pass"; it is of utmost importance that the passes overlap without marks of any kind. And it should be obvious that with a wide piece-part and too soft a wheel, there is a distinct possibility that one end or side of the piece-part will be higher than another, because of wheel wear as it moves across the work.

For belts, width is not a real problem until the belt becomes wider than 48 inches, the usual width of a trimmed jumbo roll of product. Probably most belt grinders use belts that are 12 inches or less in width. Incidentally, machines using 12-inch belts are very widely used. Machines using belts wider than 12 inches are often called wide-belt machines, with the width frequently determined by the width of the product being ground. One of the strengths of belt grinding, as was said before, is its ability to grind or polish wide work in one pass.

Work-Holding

Comparative methods of holding and/or conveying the work during processing should be mentioned, even though they will be discussed in more detail in connection with the various types of grinders. For cylindrical and centerless grinders, the work-holding mechanisms are the same; whether the machine uses a wheel or a belt is immaterial. However, for flat grinding with wheels, the choice is generally some kind of metallic magnetic chuck or workholder with sufficient strength to hold the piece-parts under wheel pressure. The magnetic chuck either rotates or reciprocates to being the work under the grinding wheel; and on high-production machines it is possible to design a grinder with an indexing work table with two or sometimes more chucks, so that while one chuck load is being ground, another may be unloaded and reloaded.

This machine design is made on the assumption, of course, that the grinding cycle is long enough to permit the second chuck to be unloaded and reloaded. If the grinding cycle is too short—and if production is high enough to warrant a custom machine—it is certainly possible to design a three-chuck grinder which with two operators would permit unloading, grinding, and loading cycles to be carried on simultaneously.

With belt grinders, high production is usually associated with conveyor belts that carry piece-parts from one grinding head to another, or perhaps a single-head with the conveyor to carry parts under it. With this setup, whether the machine has one head or several, all the human power needed is someone to load the parts; a tote box can be located at the end of the conveyor so that the ground parts fall into it. This applies whether there is one head or several. Under each head there is a backup roll sometimes called a "billy roll," or sometimes a pressure roll, or a platen (the functions of the three are identical) to ensure adequate pressure on the abrasive belt and the workpiece. And a variation of this is to have two heads mounted in a vertical plane with each other, one above a continuous sheet of steel or other metal, grinding and or polishing both sides of the sheet at the same time as the sheet is unrolled from one reel and rolled up on another after passing between the heads.

Nor is it uncommon to have, for example, a belt centerless grinder using a rubber-bonded abrasive wheel for regulating. And since the only function of the regulating wheel is to keep the piece-parts from spinning, it would certainly be possible to put more than one regulating wheel on a longer-than-usual spindle, in line with a wider-than-usual belt. (Two straight wheels cemented together for grinding would certainly leave marks on the work, but for regulating that should be no problem.) It might even be possible to have an automatic abrasive vibratory finishing unit into which the piece-parts could fall at the end of the conveyor, and thereby rough, finish, and deburr the piece-parts from one loading without any intermediate handling.

STORAGE OF COATED ABRASIVES

Grinding wheels are relatively unaffected by atmospheric conditions and require care in handling and storage primarily to prevent chipping through contact with other grinding wheels or with anything that might chip or crack the wheels. But for coated abrasives, the backings and sometimes the adhesives are sensitive to climatic conditions and will gain or lose moisture in accordance with the temperature and the relative humidity of their surroundings. High moisture content is particularly bad for glue-bonded products. Moisture causes the bond to soften under the heat generated by use, which in turn may allow the grain to shell from the bond and the backing before it is really worn.

Coated abrasives are manufactured to be at their best in heat and humidity most comfortable for human beings; that is 60° to 80°F, at 35 to 50 percent relative humidity. Temperatures above or below these ranges and air that is moister or drier can cause unnecessary trouble for coated abrasives.

The ideal storage room for coated abrasives would be one whose walls are all inside partitions, rather than outside walls. There should be neither steam-heated radiators nor hot-air inlets in the storage; if there must be a heat source in the room, it should be as far away from the coated abrasives as possible and shielded from them. Air conditioning is preferable in any case and, for any plant with a significant inventory of coated abrasives, virtually a requirement. Plants with smaller inventories could easily justify some kind of special cabinet or enclosure for this purpose.

SUMMARY

While coated abrasives cut chips similar to, though of course very much smaller than, those removed by cutting tools, and more similar to those cut by grinding wheels, there are important factors that make coated abrasives, particularly belts, worth considering for many more than their current applications. A cutting tool stays in continuous contact with the piece-part; abrasive grain in a grinding wheel cuts the piece-part intermittently, although with very short intervals between cuts; but a coated abrasive belt, with the same kind of intermittent cut, has much longer intervals between cuts—two or three or more times longer, depending on the length of the belt. This means that the grain in a belt has much more efficient air cooling than does the grain in a wheel. And the use of a coolant on a belt improves the cooling action noticeably.

A cutting tool in one pass across a piece-part cuts a thick chip that is only a fraction of an inch wide. It needs periodic resharpening and adjustment to maintain a flat surface or a consistent cylindrical surface. A grinding wheel cuts a myriad of chips from an area of a piece-part that is approximately as wide as the wheel's grinding face; and particularly if the specification is on the soft side, a wheel needs to be down-fed (or in-fed) to maintain a consistent straight surface. On a wide flat surface, if the wheel wears and there is no downfeed, the part of the surface cut last will be higher than that cut first. Similarly, on a cylinder where there is wheel wear and no adjusting infeed, the end of the cylinder cut last will be larger in diameter than what was cut first. But on the other hand, an abrasive belt will cut its width across the piece-part without significant alteration, because most of the time the belt will be wide enough to cover the whole surface of the work in one pass. And even if it needs more than one pass to cover the surface, the wear is miniscule at best and not unsuitable for most grinding work.

And of course when you consider the amount of grinding work—or metal-cutting work—that has fairly liberal tolerances for dimensions and other elements of its geometry, it's obvious that a great deal of this could be done on coated abrasive belt machines.

Furthermore, when you consider that a belt grinder equipped with a conveyor belt to carry the parts can be loaded manually and unloaded, if they are suitable, by letting them off the other end of the conveyor, there could be a considerable drop in manning requirements. And if the parts can come by chute or similar conveyance from another operation, the labor requirements drop off even more. It does not matter whether in the course of the grinding operation the piece-parts pass under one, two, or even five or six grinding heads. The loading and unloading steps in the process remain the same.

Abrasive belts have suffered even more from misconceptions than grinding wheels. The first belts, like the first wheels, were short-lived, light-duty tools. Unlike wheels, however, belts for many years were not as durable and tough as they should have been; and because lightly-built machines were sufficiently strong for woodworking, where belts were mostly used, there were only rare opportunities to demonstrate belts as metal-cutting tools. As one authority said some years back, "Before World War II, I would have said there were very few areas where belts could compete with wheels. Now I hold that there are only a few areas where they can *not* compete."

5
Flat-Surface Grinding Machines

As was stated in the preface, it is logical and convenient for management to group abrasive machining operations by the geometry of the surfaces they are designed to work on, rather than by, say, stock-removal capacity or the design of the machine. Thus, surface grinders (Fig. 5-1), disc grinders (Fig. 5-2) and free-abrasive or flat lapping machines (Fig. 5-3) will be discussed together in this chapter, even though most other publications on the subject discuss them separately.

From a technical standpoint the customary breakdown makes sense. The specification of abrasives for a wheel surface grinder, a disc grinder, and a coated abrasive surface grinder, for example, could be considered as three different areas of specialization. But specification of abrasives, beyond a very elementary level, is not an objective of this book. It has to be done, of course, but rarely by those to whom this book is addressed.

The broad use of the term "grinding" may also be questioned. It is hardly consistent with the title of this book, nor is it usually applied to fine finishing operations like lapping and honing. However, it has a broader application to surface generating and refining operations with abrasives than any other, and it is for that reason that it is used.

One observation on terminology. For many years the term "lapping" has been applied indiscriminately to any operation involving the use of abrasive grain in oil or some other carrier between two surfaces, one of which is to be finished, regardless of the geometry of the surfaces. It ought logically to be confined to flat surfaces. It has been so restricted here.

Fig. 5–1. Small surface grinders like this one are the workhorses of many shops. It has been said that 75 percent of all flat grinding could be done on a machine no bigger than this, not that that would necessarily be the best way to do it. (*DoALL Company.*)

GRINDING FLAT SURFACES

Flat surfaces may be machined—that is, either generated or refined—by abrasive processes ranging from surface grinding with wheels or coated abrasive belts to single- or double-disc grinders using discs of bonded abrasive to lapping or free-abrasive machining with loose abrasive grain. There are also conveyor-type abrasive belt machines, originally developed for polishing stainless steel and aluminum.

Range of Applications

Although all these abrasive machines work generally within limits that can be matched through considerable care only by other machining methods, they vary widely among themselves in, for instance, the amount or rate of

Fig. 5-2. Double-disc grinder with rotary carrier, probably the fastest way to grind parts with parallel flat sides. Parts are loaded just under the arm, are guided by the flat side retainers until they reach the discs, are ground, and then dropped by gravity into the box after they emerge again. (*Gardner Machine Co., Litton Industries.*)

stock removal, the dimensional precision, or the flatness or parallelism of which they are capable. There are significant differences in machine size and appearance; someone not acquainted with such machines might well conclude that they were not at all related in function or any other way. In fact, their only common characteristic lies in the fact that the work is supported on a flat surface (though sometimes with blocking to make the surface to be ground flat and parallel with the work-holder plate). There is naturally considerable overlap in capabilities, which can make the selection of the best machine for a particular application something of a problem. One key consideration is the degree of specialization that the situation will justify.

Labor costs are rarely a problem in production machining. Most require at most only semiskilled labor.

Fig. 5-3. In free-abrasive machining, operator releases pressure, slides out finished parts with left hand, and is ready to slide second retainer ring of preloaded parts onto ring with right hand. Loading and unloading are done while machine is grinding. With different tooling, this unit becomes a lapping machine. (*Speedfam Corporation.*)

Work Holding

Nor is work holding a real problem in the machining of flat surfaces with abrasives. Clamping is needed only occasionally, and fixtures are rarely complicated. The magnetic chuck used for holding practically all work on surface grinders is an outstanding laborsaver. All that is usually necessary is to place the part or parts (if parts are small they may virtually cover the chuck for batch grinding) on the chuck and then to start grinding. Of course, the chuck must be kept clean and free from dirt, swarf (mixed dirt, bits of metal and coolant), and nicks or scratches. But there is no need to clamp or otherwise restrain most ferrous metal parts. Nonferrous metals may be held by steel blocks or bars, or by specially designed blocking. But the loading-unloading cost is well below that of, say, milling or planing, where the work usually has to be clamped to the table. Obviously this takes more time and care than does simply placing piece-parts on a chuck.

Power

Power is another area of comparison which is sometimes a critical factor, and the comparative factor is what engineers term the unit horsepower. One unit horsepower at the cutter—grinding wheel, belt, turning tool, or milling cutter—is the power required to cut away one cubic inch of the material per minute. For a grinding wheel this is generally greater than for, say, a milling cutter. But if the stated figure is for a sharp cutter, many experts feel that for practical use, which often involves the use of less-than-sharp tools, the nominal figure should be doubled. It sounds impressive to say, as one standard engineering handbook does, that the unit horsepower for grinding gray cast iron is 4 to 16, while for sharp single-point tools it is 0.3 to 0.5, but when you consider that one must double the cutting-tool figure because of dulling of tools, that grinding may well require the removal of considerably less stock than is required for the cutting tool, and that for most operations the power factor is probably a minor one at best, the statement loses considerable impact. Any cutting tool requires a considerable "overcoat" of extra stock so that the cutting tool can get under the surface scale. Abrasives cut as easily through the scale as through the base metal.

WHEEL SURFACE GRINDERS

Wheel surface grinders are the largest group of machines for grinding flats with abrasives. Depending on the design of the machine, grinding may be done on the side (sometimes called the rim) of the wheel, or on the periphery. The work is usually held on a magnetic chuck or work table which functions without clamps or other mechanical restraints. The chuck may rotate or reciprocate. It may of course be loaded with one or several parts, depending on the size relationship. The spindle holding the wheel may be either horizontal or vertical. If the spindle is horizontal, the wheel usually grinds on its periphery; if the spindle is vertical, grinding is always done with the side or rim of the wheel.

Horizontal-Spindle, Reciprocating-Table Grinders

The most numerous surface grinders are the horizontal-spindle, reciprocating-table type, grinding with the periphery of the wheel (Fig. 5-4). The piece-part(s) held on the magnetic chuck (which is mounted on the work table) are reciprocated back and forth under the periphery of the wheel. The wheel can be moved up and down, and the table can be moved

Fig. 5–4.   Here the flat lands, not the V-shaped grooves, are being ground. However, with the face of the wheel dressed to the proper vee angle, the grooves could be ground or, assuming that the wheel is thick enough, a groove and a land could be ground simultaneously. (*Hitch-cock Publishing Co.*)

toward or away from the operator so that the piece-part can be ground in overlapping passes. The process is something like mowing a lawn. The vertical range of the wheel is long enough for the wheel to reach the chuck, and the traverse (in-out) movement of the chuck is enough so that its whole surface may be covered.

Most reciprocating surface grinder chucks are rectangular, although some are square, and the dimensions of the chuck indicate the capacity of the machine. For example, a common small size is called a 6-inch × 12-inch surface grinder; it can be comfortably installed in about a square yard or a little more of floor space. Its versatility is tremendous. Even though it is a small machine, estimates show that about three-quarters of all surface grinding could be done on one of these—not necessarily most efficiently, but possible.

At the other end of the scale are huge grinders which seem to stretch for half a city block across the production floor. When the work travels, the machine bed must be long enough to support nearly twice the length of the piece-parts. The length and the weight of the work set the maximum machine size. However, building a grinder with a traveling wheelhead reduces the length of machine required for a given piece-part length, a design variation used also on other reciprocating-table grinders.

On the horizontal-spindle, reciprocating-table surface grinder, which is sometimes called a type I grinder, the grinding plane is always parallel to the surface of the magnetic chuck. However, grinding work whose sides are not parallel—wedge-shaped parts, for example—is only a matter of using a tilted magnetic vise for work holding; it is merely a matter of simple fixturing. (Fig. 5-5.)

The surface pattern on these grinders is one of straight-line parallel scratches. Profile or form grinding can be readily done by traversing the wheel (Figure 5-6) if the form can be dressed (in reverse) into the periphery of the grinding wheel and runs parallel to the reciprocating motion of the

Fig. 5-5.   A part need not have parallel flat sides to be surface ground. With an adjustable auxiliary magnetic chuck, wedge-shaped parts like these can be ground with no problems. (*DoALL Company.*)

Fig. 5-6. Basic form surface grinding: slots on a flat surface. It is possible either to use a wheel of the same thickness as the width of the slot or to dress a wider wheel down to that width, or even, as a last resort, to make several passes with a narrower wheel. (*Hitchcock Publishing Co.*)

work. From an end view, the silhouette of the form is at right angles to the line of work travel. Given a grinder with sufficient power, form grinding of this kind is one of the prime advantages of the type I surface grinder, because a large number of parts with teeth or slots can be finished from the solid—bar or other basic form—often in one operation.

This type I surface grinder combines excellent stock removal with great accuracy and good finish. Probably overall it is less efficient than the vertical-spindle grinder in stock removal because of the latter's tremendous advantage in terms of wheel-piece-part contact area, but with sufficient power it does very well indeed, as witness the ability of one 125-horsepower surface grinder that removes gray iron at a rate of 40 cubic inches a minute. One big job that is often done on this type of surface grinder is the finishing of machine ways to tolerances of 0.0002 inch for dimensions, straightness, and parallelism over the length of the way, which could be several feet. Over shorter piece-parts, the tolerances could be in millionths.

The grinding of machine ways avoids the necessity of hand scraping them—and hand scraping is an art that is fading out very rapidly, because it is a hard, time-consuming, and monotonous operation which does not appeal to present-day workers. More important, of course, is the fact that surface grinding does a better-quality job much faster.

Another advantage of the surface grinder is that the work surface to be ground does not have to be continuous, as it does for milling, planing, and other cutting-tool operations. Grinding broken surfaces to the same plane can be done quite readily; in fact, such surface are probably more readily ground than are broad continuous areas. This is also a matter to be considered later when discussing the cost advantages of grinding.

Profile grinding, or form grinding. The terms profile grinding and form grinding are used interchangeably to describe the forming of an other-than-flat surface on a piece-part. For peripheral surface grinders with reciprocating tables, there are the following restrictions:

1. The form-ground area is not normally wider than the wheel face. Of course, if the form is a repetitive one, then one set of grooves can be ground, the wheel can be cross-fed by approximately the width of the form; another set of grooves can be ground, and so on, within the cross-feeding capacity of the machine. (Fig. 5–6).

2. The ridges and valleys, as well as the flats and contoured parts of the form, must run in the same direction as the wheel travels, and all widths of the valleys must be the same or narrower than the top. The valleys can be straight-sided with rounded of square bottoms, V-shaped, U-shaped, and so on. The ridges can be flat or rounded, or come to a relatively sharp edge at the top. (In practice there must be a very small radius at the bottom of the V or at the top of the sharp ridges.)

3. The length of the form is limited only by the capacity of the machine. Indeed it is often possible to line up several parts in tandem on the magnetic chuck and grind them as if they were one, particularly if the piece-parts have square ends. Such a technique reduces per-part setup time.

4. The depth of the valleys (or the height of the ridges) that it is possible to grind varies somewhat with the width of the valley or slot. A V-shaped slot (in cross section) can be deeper for a given width at its top than a U-shaped or square slot. However, within these limits there are many possibilities of form grinding from the solid.

5. As a sort of corollary to item 3, if the slots run across the width of the piece-parts, it is both possible and feasible to line them up for grinding so that several parts may be processed at one time. This may involve

some kind of fixture for holding the piece-parts to keep them correctly aligned, but it is definitely a way of increasing output.

Methods of Form Dressing the Wheel.   Form surface grinding can probably best be defined as any change in the contour of a major surface that is essentially flat—a change such as a bevel, a slot, or several slots; a surface other than one at a 90° angle to the adjoining sides; and, of course, any kind of cross-sectional contour, as viewed from the end of the workpiece.

On a reciprocating-table surface grinder, the wheel is limited to an up-and-down straight-line motion. The table of the grinder can be moved back and forth or toward or away from the operator, also in straight lines. Thus, any angular change from this level, parallel motion must be accomplished either by dressing the face of the wheel or by using an extra magnetic chuck which will hold the piece-part at such an angle that the surface to be ground will be parallel to the surface of the magnetic chuck of the grinder. Such fixtures are called magnetic sine chucks (see Fig. 5–5).

If the form involves only a simple radius, either concave or convex (Figs. 5–7, 5–8), there are numerous dressers on the market that will produce ac-

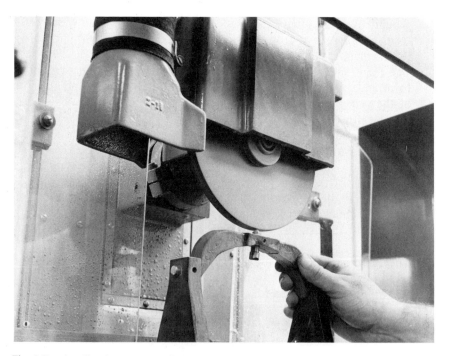

Fig. 5–7.   A radius dresser set up like this one will dress a convex radius on the wheel face and generate a concave form. (*Harig Mfg. Corp.*)

Fig. 5-8.   With this setup, the radius in the wheel is concave but the form generated is convex, just the opposite of the previous photo. (*Bay State Abrasives, Dresser Industries.*)

curate radii. Sometimes a skilled grinder operator can produce an adequate form on a wheel by hand, but it is not something to be used if the number of parts to be produced are more than one dressing will grind.

But for many forms there are thus probably two alternatives, crush form grinding and diamond form grinding.

Crush form grinding, an older technique revived about the time of World War II, involves two rolls of very hard steel shaped in the form of the finished part and mounted on a very sturdy grinder. It is for vitrified-bonded wheels only. The dressing operation consists of feeding the wheel, running at well below operating speed, into the work roll with such force that the form—or rather its reverse image—is literally crushed into the wheel. When the work roll is worn out of tolerance, the second, or master, roll is moved under the wheel to recrush the form into the wheel, which is then used, at operating speed, to regrind the work roll. The technique is interesting, one that can also be used on specially constructed cylindrical grinders. These will be discussed further along in Chapter 6.

Fig. 5-9. An assortment of diamond form dressers. (*Engis Corporation.*)

Another technique is to use a diamond-plated form block whose top has the contour of the finished part (Fig. 5-9). Dressing the wheel is simply a matter of positioning the form block on the magnetic chuck, running the wheel at the recommended dressing speed while the block is reciproacted back and forth to cut the form into the wheel, and then replacing the block with parts to be ground to form.

There is no generalization about these two approaches on which a decision between them can be made. Neither dresser is inexpensive, so either one requires a sizeable number of parts to justify itself. (There usually has to be a set of crush rolls or a diamond block or roll for each part.) The crush-formed wheel probably cuts a little more aggressively, because the crushing action leaves more jagged points on the abrasives. The diamond-cut abrasive probably gives a slightly better finish. But there is one message that must be made loud and clear: given the proper conditions—and there are many parts that fit them—form surface grinding can be a most efficient, productive, and low-cost operation. Some examples are described in Chapter 11.

Creep Feed Grinding. So far we have considered all grinding as consisting of numerous passes across—or around—the workpiece, with downfeed (in surface grinding) or infeed (in cylindrical grinding) of only a few thousandths of an inch per pass, at relatively fast work speed.

However, there is an alternative type of surface form grinding, called creep feed (or creep traverse) grinding, which differs from the norm in two ways, For one, the form-trued wheel is set to grind the full depth of the form in one pass; and for another, the travel of the piece-part is very slow indeed. It follows, then, that the wheel will most often be in full contact with the piece-part, except for a very short distance at the beginning and again at the end of the cut.

Creep feed grinding requires either a specially built grinder, or equipment which is optional on a few makes of standard machines. The reciprocal table movement (in this technique, only one way, really) must have variable speed starting at an extremely low rate, but with the ability to advance smoothly over a fairly wide range to suit a variety of operating conditions. A typical range might be from less than half an inch to about 80 inches per minute. The table must move smoothly and freely at the slow speeds, probably on roller or ball bearings, and there must be an ample supply of coolant delivered at high pressures. And the coolant system, whether for the individual machine or for a whole department or shop, must have filters capable of removing the very fine chips which are produced, because creep feed surface grinding is generally done with medium- to fine-grit wheels.

Not many machine builders make creep feed grinders; in fact, there may well be only one that does. However, when creep feed grinding is done with the proper kind of work, it can be very efficient, combining a high rate of stock removal with good surface finish and a relatively low rate of wheel wear.

Auxiliary Equipment.   While the smallish reciprocal surface grinders— say 6″ × 12″ or slightly larger—are not often regarded as high-production machines, except in the making of miniature piece-parts, it is worthwhile to mention that there are optional units—usually not made by the original machine builder—which are available for doing other forms of grinding. One of these is a rotary chuck (Fig. 5–10), which makes the machine temporarily a horizontal-spindle, rotary-table grinder that does completely different forms of work (Fig. 5–10) from that for which the grinder was originally designed. The rotating table can also index as well as rotate, for work that needs to be ground at some angle other than parallel to the front edge of the basic chuck. (This can also be done on the basic magnetic chuck.)

There is also a center-type cylindrical attachment which is essentially a work holder with a headstock and a tailstock and provision for rotating the

Fig. 5-10.   Attachments widen the range of work of small surface grinders. A rotary chuck essentially converts this grinding into a horizontal-spindle, rotary-table grinder. (*M & M Precision Systems.*)

piece-part (Fig. 5-11). When this is mounted crosswise on the magnetic chuck, the crossfeed of the grinder enables the wheel to traverse end to end of the workpiece. And, of course, if a flat is required anywhere on the essentially cylindrical piece-part, all one need do is to stop the rotation of the work and reciprocate the table as with any surface grinding (Fig. 5-11).

There are also available a high-speed spindle for incidental internal grinding, a vacuum chuck which holds the work by exhausting the air from beneath it (often recommended for very thin work), and an array of wheel dressers for generating wheel grinding faces other than the standard flat face at right angles to the sides. For the latter there is usually a dresser built into the machine, or there are inexpensive tools that can be mounted on the machine's magnetic chuck.

Vertical-Spindle, Rotary-Table Surface Grinders

This is a high-production high-stock-removal machine, known in shops as a vertical grinder, or sometimes simply as a surface grinder. For a given

Fig. 5-11. This adjustable attachment converts a small surface grinder into a cylindrical grinder. As shown, it is set up for taper grinding. (*DoALL Company.*)

horsepower rating and material, this type removes more stock than any other grinder, and for this reason it has been outstanding in abrasive machining. In fact, most of the early examples of the use of grinding wheels for heavy stock removal as well as finishing were work done on vertical-spindle, rotary-table grinders (Fig. 5-12).

Grinding is done with either a cylinder, a cup, or a segmental wheel, which means that the process involves a flat abrasive cutting surface grinding a flat piece-part surface. The comparative size of the resulting contact area ensures fast stock removal.

The second distinguishing feature of this grinder is that the piece-parts—or sometimes a single piece-part—are mounted on a rotating magnetic chuck (Fig. 5-12). It is obvious that those pieces mounted toward the outer rim of the chuck travel at a greater speed than those placed nearer the center, so that a piece-part placed very close to the center would travel at a very low speed, if indeed it traveled at all. There is always an open space in the middle of the chuck or table. This variation in speed also produces some problems in determining work speed, inasmuch as a part on the

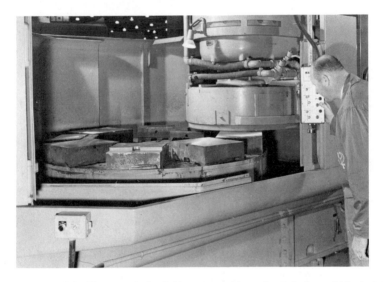

Fig. 5-12. One of the bigger vertical-spindle, rotary-table surface grinders, which started the abrasive machining trend. (*Mattison Machine Works.*)

outer rim may be traveling three to six or more times as fast as one toward the center.

Vertical-spindle surface grinders tend to be big-production machines. There is not the range on the small side that there is in, say, horizontal-spindle, reciprocating-table grinders, although such grinders are of course quite possible.

But even though the grinders are big, they are capable of very precise dimensional accuracy, extreme flatness, and superior surface finish, all in one setup. For example, it is common to mount a load of piece-parts on the magnetic chuck and, with the wheel dressed for stock removal, remove excess material at a fast rate and then, without changing the piece-parts on the chuck, re-dress the abrasive wheel for finishing and complete the machining of the parts to the required dimensions, flatness, and quality of surface finish.

However, it must not be concluded that this is a universal grinding method for producing flat surfaces. The working abrasive surface of the wheel is parallel to the chuck or table surface, and any part to be ground by this method must have the geometry to permit it to be mounted with a substantial surface in contact with the chuck (for holding power) and with the surface to be worked parallel with the chuck and the wheel face. And since this grinding method can involve considerable pressure, it is not usually suitable for delicate piece-parts.

But there are still hundreds, even thousands, of piece-parts for which this

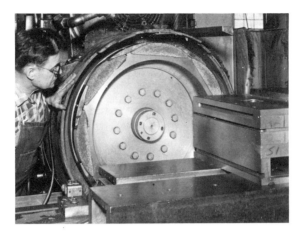

Fig. 5-13. This modification of the surface grinder is called a face grinder, with segments in a chuck mounted on a horizontal spindle; it is used for piece-parts which because of their geometry are more easily machined with the work surface in a vertical position. This machine has a traveling table and a stationary wheelhead. Face grinders are usually large, and may be as much as 20 feet in length. (*Hitchcock Publishing Co.*)

is a preferred method, provided also that there is sufficient volume of production to justify the cost of the machine. And where production is very high, as in the manufacture of automobiles, for example, it is quite feasible to design vertical surface grinders with two or sometimes more indexable tables, so that while one table load is being ground, the other(s) can be unloaded and reloaded. With a duplex table design, the grinding wheel is working nearly all the time, so the operator has very little watching time.

Two other elements of chuck loading ought to be mentioned. Mentioned previously was the need for keeping the surface to be finished parallel to the chuck surface and the grinding surface. Such parallelism can also be achieved by using appropriately shaped iron or steel fixtures which will hold the piece-parts in proper position. The technique can be adapted for, say, a section of a steel housing which has to have a flat rim so that it may be joined to another mating part, even though the housing itself is rounded. Some such fixtures require a substantial production level to justify their cost, for if the production level is sufficient, the savings can be significant. Quite often, though, fixtures are not costly.

The second element of loading is the use of steel retainer rings or blocking. Retainer rings, which are simply steel rings of appropriate diameter, are most often used for small parts which might otherwise be swept off the chuck by the pressure of the wheel. Steel blocking, in this as in practically all forms of surface grinding, helps to keep nonferrous parts on the chuck for grinding. Blocking might also be used when the piece-parts are elongated and the ends are to be ground. Then appropriate blocking helps to keep the parts upright and in proper position for grinding.

When a piece-part can be most conveniently ground on this type of surface grinder, it is almost always possible to improvise the necessary blocking or retainers to enable the grinding to be completed.

If what has just been said creates the impression that workholding on the magnetic chuck of any surface grinder is rarely a real problem, that is just about the truth. And the use of double-faced adhesive tape, similar to that used for carpets, has not been mentioned. But it's a practical way to hold thin, especially nonmagnetic, workpieces. And clamping is as practical as it is with any milling or other cutting-tool machining operation but it is required not for every part, as in milling, but only where the geometry of the part provides insufficient holding power on the chuck or table.

There is probably considerable lateral force exerted on the surface to be ground on a vertical-spindle surface grinder, because of the large interface between the grinding wheel and the work, more than there is on any type of peripheral surface grinder. But even at that, the lateral force on the grinder is much less than it would be for cutting-tool machining.

Gaging the Finished Work. Accurate measurement of the finished parts is of course essential, perhaps more so in abrasive operations than in any other. And in this type of surface grinding, which is often a volume operation, it becomes even more important.

Of course, if the grinder is fully automatic, there is no problem. When the parts are ground to size, the machine stops and the operator unloads and reloads the chuck. Direct measurement of the parts is also a possibility. But with rotary surface grinding, where the chuck is usually loaded with a number of parts, there is another very simple and inexpensive way to check. All that is required is to include in each chuck load one finished part whose top is coated with prussian blue or some other similar material, and when the wheel begins to cut that part, the load is finished. This degree of accuracy is sufficient for a great deal of production work, though it might not be adequate for very close work. This is one of the small advantages of grinding that give it an edge over competitive methods of machining.

The two types of surface grinders described above—the horizontal-spindle, reciprocating table type on which there is a peripheral grinding action and the vertical-spindle, rotating-table type on which grinding is done with essentially a flat abrasive surface—make up the bulk of all surface grinders.

The rotary-table grinder is capable of high production and considerable accuracy on many flat parts. It can also do a considerable amount of other flat machining. Moreover, as should be obvious, the capabilities of the two types of surface grinders often overlap, a situation which can be beneficial when there is a varied flow of work. In a general discussion such as this, it is not feasible to spell out in detail what one must consider before purchasing any particular type or size of grinder(s). The point is to give readers an understanding of the capabilities of the various types, so that they are aware of what the grinders can do, broadly speaking. The specifics of types, sizes, and similar details are left to those who are specifically responsible for these areas.

Specialized Surface Grinders

Grouping other variations of flat-surface wheel grinders as special is somewhat arbitrary, but the fact is, these other combinations of grinding action and table or chuck conformation do not have the broad variety of applications that is characteristic of the two types just discussed. So, whereas almost any machining shop, department, or section will have two or more reciprocating-table surface grinders and at least one with a rotary table, in sizes appropriate for the facility's production, their having a grinder with a horizontal spindle and a rotary table would be less likely, and one with a vertical spindle and a reciprocating table, even rarer. And there is even a type, traditionally called a face grinder, on which the piece-parts are mounted so that the surface to be ground is vertical, and the grinding is done on the flat of segments (Fig. 5-13) mounted on a horizontal spindle. In fact, there are variations on variations, mostly machines made by a company that are built for one type of work. It is entirely possible that such machines would be good investments under given circumstances. After all, the more closely a grinder or any other machine tool can be matched to the particular requirements of a job, the more efficiently that job will be done. How an investment can be justified depends, of course, on many factors. Even in automotive manufacturing, where it is not uncommon to set up a machine to run one operation on one part throughout an entire model year—which is about as specialized as you can get—consideration has to be given to other uses of the machine when the model year is finished.

Horizontal-Spindle, Rotary-Table Wheel Surface Grinders.   In contrast to the vertical types, these grinders (Fig. 5-14) are frequently used where extreme flatness is required in an essentially round piece, and/or where a circular scratch pattern on the piece-part (resulting from mounting one part centered on a rotating table) is desirable. These machines are not considered capable of as high-stock removal as are other types of surface

Fig. 5-14.   Surface grinder with horizontal-spindle, rotary table, grinds ring with a concentric scratch pattern. (*Cincinnati Milacron, Heald Machine Division.*)

grinders, but their accuracy and the surface finish they can produce with fine-grit wheels make them worth consideration. Finally, many grinders of this class are equipped with tilting tables that provide an unusual and efficient means of grinding both concave and convex surfaces (Fig. 5-15). One such example would be the hollow grinding of circular saws. Hollow grinding in this instance merely means that the outside diameter of the saw is slightly thicker than the central portion, so that there will be no rubbing of the center part of the saw on the sides of the cut. It is readily done on this type of grinder.

Vertical-Spindle, Reciprocating-Table Wheel Surface Grinders.

These are big grinders, ranging from 8 to 18 feet or more in length, using either cup, cylinder, or segmental wheels. Horsepower may go up to 150, or even more. Stock removal is high. With segmental wheels and sufficient horsepower, one of these machines could remove stock upward of our times as fast as a similar peripheral wheel grinder.

Fig. 5–15. With the worktable tilted to grind a 10° concave surface, the machine readily completes a machining job that is unique. (*Cincinnati Milacron, Heald Machine Division.*)

The wheel spindle is vertical, and the surface to be ground horizontal. For very long piece-parts, where the length of travel could present a space problem, there is a traveling wheelhead variation, which reduces the total floor space needed by the machine and the workpieces. And it is worth noting that on traveling-head machines, it is practical to utilize the machine as a dual-table grinder, grinding work on one end of the chuck while the other end is being unloaded and reloaded.

With high horsepower and segments, the vertical-spindle machines because of the broad area of contact between the abrasive and the piece-parts, can remove considerable stock, indeed, up to three or four times that of peripheral wheels. The abrasive on any vertical-spindle grinder produces a cross-hatch scratch pattern that is considered very good for sliding bearings.

The flat-to-flat contact makes this type of machine adaptable to work on an interrupted surface, one with slots or holes. The machine is a heavy producer, as has been noted, but there is considerable heat generated, which usually requires use of a coolant.

Fig. 5-16. A vertical-spindle, reciprocating-table machine is useful for grinding long, flat piece-parts to close tolerances and good-quality finish. (*Mattison Machine Works.*)

Fig. 5–16 shows some of the capabilities of the grinder, and Fig. 5–17, something of its range with different setups.

Surface Finish, Flatness, Roundness, and Parallelism

Most people tend to think in terms of surfaces that are smooth, flat, rough, round, or parallel as definite characteristics that the surface either has or has not, but in machining terms such is not exactly the case.

First, it must be said that there is no surface of any kind that is absolutely smooth, flat, parallel, or absolutely round, which will be discussed later. It must also be said that industry requirements, and the means for achieving them as well as the means for determining the degree of achievement, have all come much closer to the absolute than was the case even a few years ago. We can produce parts to tolerances that were unthinkable a couple of decades ago. Fortunately, we have simultaneously accepted the fact that the ultimate is neither always needed nor even desirable.

With respect to surface finish, for example, any machined surface, regardless of how smooth it may feel or look, consists of a pattern of scratches. Of course, in any good-quality surface these can be discovered only through magnification, but the scratches are always there, and any gaging or checking of surface finish becomes simply a matter of determining the depth and sometimes the pattern of the scratches.

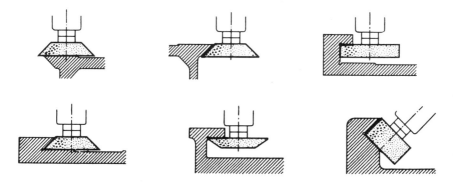

Fig. 5-17. End view shows variety of work possible with a vertical-spindle, reciprocating grinder with different wheel shapes and a tiltable spindle. (*Mattison Machine Works* and *The Grinding Wheel Institute*.)

On reciprocating surface grinders, a peripheral grinding wheel produces a pattern running essentially parallel with the direction of table travel. If the table rotates and only one piece-part is centered on the table, the scratch pattern will be one of annular parallel lines, somewhat resembling the growth rings on a tree. Any side grinding wheel produces a scratch pattern of intersecting arcs sometimes called a "dutch" finish (Fig. 5-18), and such a pattern is considered useful for bearing surfaces. The intersecting scratches are excellent for holding minute drops of oil, which is, of course, a desirable characteristic in any bearing surface.

DISC GRINDERS

When parts with two parallel sides are ground on a surface grinder, they must be turned over and ground on the second side. This is not the most efficient procedure. But when such parts are ground on a double-disc grinder, both sides are machined parallel in the one operation; and because the parts can be fed in close sequence between the flat surfaces of the discs, the machining process is very rapid. In a sense, double-disc grinding bears the same relationship to surface grinding that centerless grinding does to cylindrical grinding. The first two generate flat surfaces; the other two generate round surfaces like shafts. Disc and centerless grinding are specialized high-production operations. Once the abrasives are selected and the machines set up, the operator need only load and unload the grinder, a function that may sometimes be made automatic.

Surface and between-centers cylindrical grinding can be made more productive but can rarely reach the production levels of their counterparts. On the other hand, surface and cylindrical grinders are much more flexible and

Fig. 5–18. Finished part shows scratch pattern of intersecting arcs produced by vertical-spindle grinders. (*Hitchcock Publishing Co.*)

can handle a greater variety of work with fewer changes. So it is not surprising that centerless and disc grinders tend to be big machines concentrated in industries like the manufacture of automotive and farm machinery, where high-speed feeding (Fig. 5–19) is a necessity. A shop with half-a-dozen surface grinders of various sizes and two or three cylindrical grinders might possibly have one centerless grinder, but it would not often also have a double-disc grinder, only because it is difficult for many shops and plants to generate the kind of production for which it is best suited.

Single-disc grinding, incidentally, used to be more popular than it is today, but it has fallen out of favor, most likely for the reason that surface grinders are probably more efficient for much of the work once done on single-disc machines.

Double-Disc Advantages

Double-disc grinding on horizontal-spindle machines has the following advantages:

1. It is a high-production operation that with favorable piece-part material and geometry can be very high indeed. If there is considerable stock to

Fig. 5-19.   Closeup of rotary feeder for double-disc grinder. Note the retainer chain which keeps the parts in the feeder until they can be safely dropped into the discharge chute. (*Hitchcock Publishing Co.*)

be removed on a large part, the rate will obviously decrease. However, as you can see in the accompanying tabulation (Fig. 5-20), a pump rotor made of sintered powder metal could be ground at only 600 per hour, even though the geometry was favorable and the stock removal at most only three-thousandths. But a connecting rod of forged or cast iron can reach 1000 per hour despite an awkward shape and a stock-removal requirement ten times as great (0.030 inch). For finish grinding of piston rings, where the depth of stock removal is the same as that for the rotors (although both area and volume of metal removed are less) and where both piece-part shape and material are favorable, the production rate shoots up to 45,000 per hour.

2. Grinding is continuous on both parallel sides, while the piece-parts are between the discs, and the whole disc surface is used.

3. Once a double-disc grinder is set up, the primary functions to be performed manually are loading and unloading, both of which can quite

easily be made automatic. In fact, manual unloading is required only for very fragile parts which might be damaged by contact with others.
4. Finished parts have very good parallelism, finish, and flatness. Size (i.e., thickness) is uniform, particularly with automatic size control. Stock removal (depth) can range up to 1/4 inch or a little more.

Limitations

Disc grinding also has a few limitations, as follows:

1. Parts must be uniform in size and fairly close to finished dimensions. Actually the two discs are not exactly parallel; they are angled a little bit so that the entry-side opening is a bit wider than the exit side, permitting the piece-parts to be fed through and come out with parallel sides. The fact that all parts must pass between the discs is a limiting factor for part-thickness oversize.
2. Chip disposal and coolant application can be a problem at times, although the designers of the machines and of the abrasive discs have been able to minimize the difficulties.
3. If the parallel areas to be ground are made of different materials or if they differ substantially in area, there can be some difficulty in selection of the abrasive disc specification—the abrasive, grit size, and so on—and the grinder might be operating with a silicon carbide disc on one side and an aluminum oxide disc on the other. This could be the case if, for example, the piece-parts were bimetallic, with one side brass and the other side steel. And if one side to be ground were substantially different in area from the other, with both of the same material, then the abrasives in the discs would be the same; but there might be differences in grit size or grade, or perhaps even in bond.
4. As a generalization, the discs grind the entire exterior surfaces of the two parallel sides, and it is accepted that parts with bosses or protuberances cannot be handled. However, when one supplier was confronted with a part having a boss or raised section on one side, he was able, with the cooperation of his supplier, to work out a system using a smaller disc on one side, so that the boss did not interfere with grinding both parallel sides of the piece-part.

Other Applications

Although it is common practice to think of disc grinding in terms of parts like piston rings where the parallel sides are obvious, it should also be remembered that the process can be used to grind the heads and the stem

Customer		Workpart and Operation—Two Sides at a Time	Production Rate
Automotive Manufacturer 		Part .... Connecting rod Size .... 8 1/2" (215.9 mm) overall length Material ...... Forged or cast iron Stock removal 0.030" (0.76 mm) overall, 1 cut Parallelism ... 0.001" (0.025 mm) Uniformity .... 0.002" (0.051 mm)	1,000 per hour
Pump Manufacturer 		Part .... Pump rotor Size .... 2.120" O.D. (53.8 mm) Material ...... Sintered powder metal Stock removal .0.0015" to 0.003" overall (0.038 mm to 0.076 mm) Flatness ...... 0.002" (0.0051 mm) Parallelism ... 0.0002" (0.0051 mm) Uniformity .... 0.00035" (0.0089 mm)	600 per hour
Piston Ring Manufacturer 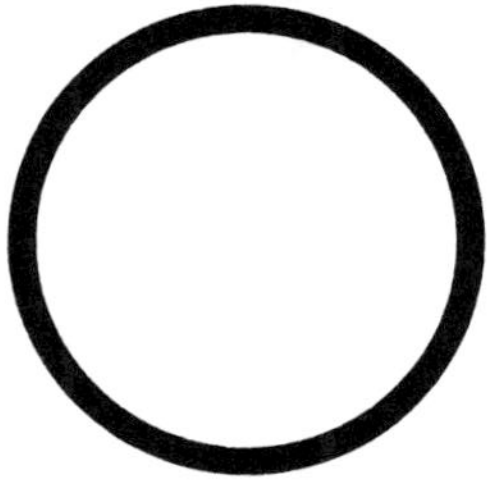		Part .... Piston ring Size .... 2 5/16" O.D. x 2 1/16" I.D. (58.7 mm x 52.4 mm) Material ...... Cast iron Stock removal .0.010" to 0.013" (0.25 mm to 0.33 mm), rough grind 0.002" to 0.003" (0.051 mm to 0.076 mm), finish grind Parallelism ... 0.0003" (0.0076 mm), rough grind 0.0002" (0.0051 mm), finish grind Uniformity .... 0.0005" (0.013 mm), rough grind 0.0003" (0.0076 mm), finish grind	35,000 per hour, rough 45,000 per hour, finish

Fig. 5-20. Three parts processed on double-disc grinders, with part and machining data and production rate. (*Gardner Machine Co., Litton Industries.*)

ends of gasoline motor valve stems, or the front and back surfaces of rear axle housings. Of course in both these applications, there must be specially designed work carriers, and the discs must be farther apart than they generally are. But the point is that if two parallel surfaces on a piece-part have to be finished to close tolerances for surface quality, flatness, and parallelism, disc grinding ought to be considered seriously. For some work, like the valve stems, the application of disc grinding is not so readily apparent; but once one gets past the obvious, it becomes quite clear. Disc-type machines have also been built for, say, grinding the ends and sides of cast-iron bath tubs, and for grinding bricks that need close-tolerance fitting. In the brick-grinding application, three machines are set up in sequence, with two conveyors that turn the bricks as they go along. So with just one manual loading, the first machine grinds two sides; after the brick is turned, the second machine grinds the other two sides; and then after a second turn, the third machine grinds the ends.

Work Carriers

The design of the work carrier is a critical factor, perhaps a unique one, in double-disc grinding. It has been noted that most double-disc machines have vertical grinding surfaces, a condition which tends to simplify the design of the work carriers and to facilitate coolant application and swarf disposal. So the work carrier is basically a support and retainer, rather than a rigid clamp, for the piece-parts. The carrier must perforce be thinner than the piece-part, and the openings in the carrier should be roughly similar in shape to the piece-parts, although it is not necessary that they be exactly to size. The work is retained on the bottom and top by the carrier and on the sides by the discs, so no harm is done if it slides around a bit within the carrier.

There are three general types of carriers—through-feed, rotating, and reciprocating, plus a host of specialized holders for such diverse parts as cylinder rods, valve stems, and bathtubs; but only the general types need to be discussed in any detail (Fig. 5–21).

The simplest of the carriers—and the most productive for appropriate parts—is the through-feed type, which is made up of a bottom track and a top track, plus side guides that go up to and away from the discs. Such a carrier can be inclined so that round parts—piston rings are a good example—will roll down the track, between the discs to be ground, and then out the other side, either to temporary storage or to whatever is the next step in processing. Provided that adequate infeed storage can be worked out, such a unit requires only casual monitoring by an operator, particularly if the gaging setup stops the grinder when the parts begin coming out oversize.

Feed Thru

This type of fixturing provides maximum production. Power driven belts, lug chains, or rolls feed the parts between the opposed discs, while upper and lower guide bars support the parts through the grinding area.

Rotary Carrier

This method offers the next highest production rate. The parts are traversed in an arc between the opposed discs. Parts may be loaded and unloaded manually or automatically.

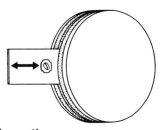

Reciprocating

Reciprocating fixturing is recommended for more refined tolerances, for lower production requirements, or for larger parts not suitable for feed thru or rotary methods. Held in the fixture, parts are reciprocated between and across the disc surfaces.

Fig. 5-21. Three types of double-disc work carriers. (*Gardner Machine Co., Litton Industries.*)

Given the appropriate parts, as noted, this can be an incredibly fast operation, because there are no intervals between the parts in process. For finish grinding of piston rings, as an example, where only a small amount of stock must be removed from each ring and the area is also small, a production rate of 40,000 to 45,000 pieces per hour is quite possible.

A second common type of feeder is the rotary type, which is usually a plain disc of steel with appropriately shaped openings around its outer diameter to retain the parts. The disc is mounted so that as it rotates, the parts are carried between the discs in an arc. Usually the parts must be loaded near the top of the arc, but at the bottom, after they leave the discs, parts fall out from their own weight onto a conveyor or into a tote box.

A variation of this type is an indexing feeder with three openings, usually for larger parts. In operation, one opening is in the loading position, the second is in the grinding position, and the third is in the unloading position. After a part is processed, the holder indexes, unloading the finished part, swinging a new part in for grinding, and presenting an empty holder to be loaded.

Obviously, no one of these feeders is as fast as the through-feed type, partially because of the design and partially because these are used for larger parts, where the stock to be removed often is both larger in quantity and more difficult to remove. But this is only in comparison with the faster feeder; rotary feeders often have a production rate of say 600 to 1000 or more.

The third general type is a swinging arm which oscillates one part at a time between the discs. It has an opening similar to that of the rotary feeder, and it is used for heavier parts. Since the load and unload steps have to be done each time the carrier swings out from between the discs, this feeder is slower than either of the other two, but it is almost certainly faster and more precise than any other alternative way of machining the parts.

Needless to say, these are not the only feeders available; one or another of the suppliers of disc grinding machines can usually design something for practically any piece-part that requires the finishing of two parallel surfaces, whether these are a fraction of an inch apart or literally a few feet apart.

Abrasives for Disc Grinding

Abrasives for disc grinding follow the same general patterns as for other types of grinding, but the geometry of the tools is different, to fit the application. Smaller discs are generally molded in one piece, with outside diameters ranging from 10 to 48 inches, and thicknesses from 1 inch to 3 inches. Discs are generally thinner than most grinding wheels. The center portion of the disc is not a very effective grinding area; similar to rotary surface grinding, the abrasive does not move fast enough to be efficient, so the center is often open, with the parts being ground definitely above or below center. Discs from 48 to 84 inches in diameter are generally molded in segments like pieces of pie. All discs are attached to backing plates; either bolt holes are molded into the disc or nuts are molded or cemented into the disc. The point is to have positive mechanical holding of the disc on the back plate. Coated abrasive discs can be cemented onto a back plate (indeed, this was the first kind of disc used), but these are not in general industrial production use.

Coated Abrasive Surface Grinders

The generation of flat surfaces by means of coated abrasives—generally belts—ought logically and commonly to be described as coated abrasive surface grinding. But such a statement is only half right; the description is logical but not common.

Theoretically, of course, it would be quite possible to remove the wheelhead assembly of almost any bonded abrasive wheel surface grinder with a horizontal spindle and replace it with an abrasive belt assembly, including the contact wheel or roll and the backstand. The reciprocating or rotating magnetic chuck and the mechanism for raising and lowering the abrasive tool need not be changed, and the mechanism for crossfeeding would possibly not be needed, because the belt could be adjusted to the reciprocating chuck (or the rotating chuck). However, it would probably be better to design the whole machine from scratch.

There is also the long-established conveyor-type of coated abrasive surface grinder (see Fig. 1-19) that might be described by another name such as sheet polisher, because one of the traditional uses for such a machine is to polish stainless steel and other types of steel sheet and another, to finish plywood in 4- × 8-foot sheets, taking advantage of the fact that coated abrasive belts can be made 4 feet wide with only one diagonal splice.

For a proper evaluation of coated abrasive belt surface grinding, it is necessary to recapitulate some of the advantages of these tools versus bonded abrasive wheels and cutting tools.

Any abrasive, either a wheel or a belt, is essentially a cutting tool with millions of points rather than one or just a few. Cutting tools generally remove more stock per pass, although not necessarily per minute or shift, but the real champions in stock removal among abrasive machines are the relatively large, high-powered grinders with an essentially flat-to-flat abrasive-work contact area. When an abrasive belt becomes dull, it is replaced or moved to another, less-demanding (in terms of stock removal) station, an operation that takes at most 3 minutes and is usually performed by the operator. When an abrasive wheel becomes dull, it is dressed (sharpened) by the operator; and in automated operations a dressing phase is incorporated into the cycle, usually without downtime. For example, while one side of a wheel is machining, the opposite side can be dressed as needed. Both wheels and belts can provide closer tolerances and better-quality finishes than any cutting tool, and at good production rates. Generally the power demand for any grinding is greater than that for cutting-tool machining, but this is usually offset by the self-sharpening (or at most operator sharpening) that abrasives require, set against the demands of cutting-tool resharpening, which requires skilled operators and, in many shops, special equipment.

Stock Removal.   Stock removal rates on any type of surface grinder will change according to the work material. Belt grinders have been used primarily for aluminum and chrome steel, as well as for gray, cast, malleable, and ductile irons; on the latter metals, stock removal reaches 30 to 60 cubic inches per minute *per square inch of belt contact* (Fig. 5-22). The parts include harrow discs, high-carbon steel farm implement blades, forgings, castings, and aluminum parts, all with flat surfaces. Because the belt is able to cover the whole area of the piece-part at once, the per-square-inch (or unit) pressure is low, so it is possible to grind and finish parts where the interruptions add up to a surface area greater than that to be ground.

When one considers the amount of stock removed, tolerances are very attractive. Dimensions and parallelism can be held to about ±0.002 inch; flatness from 0.002 inch; and surface finish anywhere from about 350 microinches—pretty rough—with a 24-grit belt, the coarsest usually used in this type of operation. With an 80-grit belt the surface finish can be improved to 125 microinches; and it can be further improved with finer belts.

Some types of platen grinders have been around for over half a century, but primarily as woodworking and home workshop tools. However, with improved belts, larger and more rugged machines, and more horsepower they have become full-fledged industrial machines. Formerly, parts were held in the operator's hands against the belt to control the cycle. Now the operator merely loads a part (or parts) and pushes a button and lets the controls take over the rest of the processing cycle, He returns the piece-part(s) to the unload-load position when it (or they) are finished.

Surface Platen Grinders.   Two of the most significant developments in abrasives during the 1970s were the development of a high-powered belt

Fig. 5-22.   This huge coated abrasive surface grinder uses two 36 × 126-inch belts with high horsepower to remove up to 30 to 60 cubic inches of stock per minute. The machine has four indexable worktables: no. 1, in front of the operator at 6 o'clock is the unload-load station; no. 2, at 9 o'clock is the rough grinding station; no. 3, 12 o'clock can be used for an intermediate operation like drilling; and no. 4, 3 o'clock is for finish grinding. (*White-Sundstrand Machine Tool, Inc.*)

surface grinder and the emergence of the platen grinder as a full-fledged industrial machine tool. Both machines are now available with higher and, in some instances, much higher horsepower operating with solid steel backups for the piece-parts and using much-improved belts that will stand the gaff. Belt surface grinders can use magnetic chucks similar in function to those used on wheel grinders, or nest-type fixtures, or sometimes simple clamping to hold the piece-parts.

The basic platen grinder consists of two pulleys (Fig. 5-23) mounted vertically, over which a belt is mounted so that it moves down over a steel or carbide plate mounted in line with the pulleys. The piece-part is pushed against the descending belt toward a work stop. Most of the former platen grinders were probably operated without a coolant, but the new industrial types use one, so that the belts can cut freely and without heat—in fact, without any distortion or stress on the piece-parts.

Stock-removal rates range from 0.001 to 0.015 inch per second; and cycle

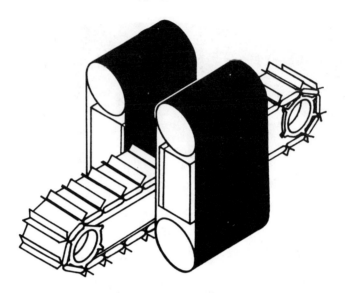

Fig. 5-23. This industrial-sized platen grinder will remove stock from flat surfaces at rates up to nearly 1 cubic inch per minute, depending on part size and material. Softer material may machine faster. Flatness tolerance ranges from 0.0005 to 0.003 inch, depending on part and tooling method. (*White-Sundstrand Machine Tool, Inc.*)

times are usually from 10 to 30 seconds. Flatness of the part, a primary benefit of this method, normally ranges from 0.0005 to 0.003 inch, depending on the part and the tooling method.

Belt Sanders and Polishers. The established traditional coated abrasive machine for finishing or deburring (Figs. 5–24, 5–25, 5–26) flat surfaces is a conveyor-fed grinder with one or more coated abrasive belt heads, with a belt (or belts) wide enough to grind material, either in lengths or individual piece-parts, across the width of the conveyor. Backup support of the work can be either a roller (called a billy roll) lined up with the belt head to exert pressure on the piece-parts as they pass under the abrasive or a flat piece of steel. Contact rolls are usually hard rubber, serrated to make them more aggressive. With a change of abrasive, some of these can become surface grinders.

If the machine has more than one abrasive head, then the same setup exists at each head. The conveyor is, of course, set up to carry the piece-parts under the heads in succession. The newest belt is always mounted on the first head of the sequence, then moved to the second, and so on for each belt change, until it is finally removed from the last head and discarded. As the abrasive on the belt wears, the belt tends to have less of a cutting action

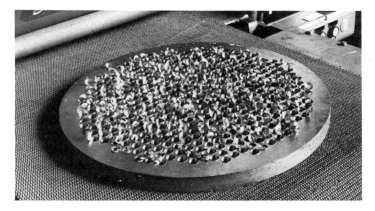

Fig. 5-24.   This part has been badly burred as a result of a drilling operation. It is ready to be deburred on the belt grinder. (*Timesavers Inc.*)

which removes stock, and more of a polishing action which improves the finish.

Lapping and Free-Abrasive Machining

Lapping and free-abrasive machining are two closely related abrasive processes for finishing flat surfaces by removing tool marks, slight distortions, wavy surfaces, and similar minor defects remaining from preceding machining operations. In fact, they are so closely related that there is considerable body of opinion that maintains they are really one. Both essentially involve abrasive action by abrasive in a vehicle (most often an oil) which rolls between the work and the rotating lapping plate under pressure. In lapping, the plate is somewhat soft, frequently cast iron, so some of the abrasive will become imbedded in the lap plate, thereby becoming an abrasive tool. Laps are also made of steel, brass, copper, and aluminum, for various purposes, but cast iron is generally the favorite. Of course, there is abrasive rolling around between the piece-parts and the lap plate, because for various reasons, not all of it becomes imbedded.

In something of a contrast, the backup plate in free-abrasive machining is hard enough so that the abrasive does not become imbedded. Thus, all the abrasive is rolled between the work and the plate for all of its useful life. (Fig. 5-27)

However, there is a distinct difference between the two methods in the sizes of abrasive used. Lapping abrasives are normally finer.

Lapping is not intended for removing any significant amount of stock; for years it has been considered as a process for ultra fine finishes, extreme

Fig. 5-25. Same part after deburring. (*Timesavers, Inc.*)

flatness, extremely close tolerances and the ultimate in parallelism. Moreover, lapping is not a fast operation.

For free-abrasive machining, however, definite stock removal is claimed, in the range of 0.005 inch to 0.018 inch per hour for cast iron or 0.005 inch to 0.050 inch for stellite. In each case the stock removal rate for the finest abrasive is stated first; the second is for the coarsest of the four popular sizes of abrasives. Surface finish (rms), in the same order, ranges from 6 to 30 for cast iron and from 8 to 16 for stellite. In general, the harder the material, the better the possible surface finish.

For other materials, soft sintered iron for instance, the stock removal rate claimed ranges from 0.010 to 0.040, and surface finish only from 14 to 35. However, on hard (60 R_c) steel, the stock removal rate is only from 0.002 to 0.008—not much—but surface finish improves to 3 and 15. Nylon is rated at 0.030 to 0.120, and 14 to 30.

So, even though the distinction may not be clearcut, there are certainly applications on which a somewhat greater stock removal rate than is customary for lapping could be useful.

Flat Lapping with Wheels and Flat Honing

There are two other specialized flat finishing operations—one called flat lapping with abrasive wheels and the other, flat honing—which are

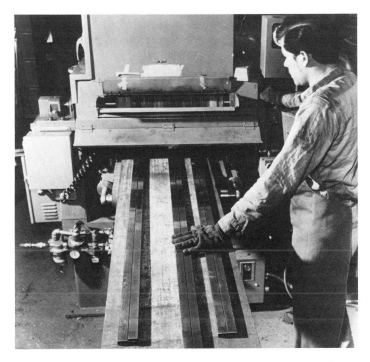

Fig. 5-26. This belt polishing or grinding machine (depending on belt used) handles several lengths of rectangular stainless tubing at one time and is reported to have increased the operator's productivity by a factor of 4. (*Hitchcock Publishing Co.*)

holdovers from the time when virtually any fine finishing operation might be termed either lapping or honing, or both.

Flat lapping with abrasive wheels—the wheel(s) replace either the top lapping plate or both plates—is in principle essentially a disc grinding operation with the discs on vertical spindles and much finer than customary bonded abrasive wheels or discs.

Flat honing produces incredibly flat surfaces-flat within one light beam or 11.6 millionths of an inch. The process is no different in principle from those which have already been discussed. Either loose abrasive grain or bonded abrasive wheels are used, depending on the type of piece-part. The piece-parts both oscillate and rotate across a rotating finishing plate, and parallelism within the same range is brought about by finishing both surfaces at the same time on a double-surface machine. But logically it must be said that when loose grain is used, the process is identical with lapping; and when bonded abrasive wheels or discs are used, it is really a single- or double-disc operation with extremely fine-grained discs.

Fig. 5-27.   This flat-surfacing machine is capable of either lapping or machining, depending on the size of the loose abrasive and the hardness of the plates. (*Speedfam Corporation*)

SUMMARY

This chapter has established the point that there are many machines and processes for generating flat surfaces on almost any material used in industry, with any desired degree of surface finish and/or parallelism. It is also apparent that there is considerable overlapping in terms and in process capabilities; and, moreover, there are no hard-and-fast rules governing the choice of machine or process. In fact, with several machines capable of doing practically any job, the choice may well depend on factor(s) other than machine and process.

The point to be stressed here is that the choice ought to be made on the basis of familiarity with a wide range of processes.

6

Cylindrical Grinding Machines

The abrasive-using machines to be discussed in this chapter have as a common characteristic rotation of the piece-part against the periphery of a grinding wheel or a coated abrasive belt on a contact roll. All the grinding is done by the outside diameter of the abrasive wheel or the belt; there is no action with the flat of any abrasive element. If the surface of the piece-part to be finished is itself an outside diameter, then the operation may be known as external cylindrical grinding; if the surface to be ground is some kind of internal diameter—not the major diameter of the part—then the operation is called internal grinding. One other point: though most piece-parts do rotate during grinding, there are some that move in an elliptical fashion to produce cams and similar parts that are generally more cylindrical than flat but not necessarily round. There are also planetary machines designed to have the abrasive belt system rotate around the work.

It is not particularly surprising that some of the first grinders produced were cylindrical grinders. When manufacturers needed better finishes and tolerances on shafts and pins than steel cutting tools would given them, some ingenious individuals removed the cutting tool from a lathe and substituted a grinding wheel. The grinding wheel may have been only 3 inches in diameter and 1/16 inch thick, but it was without doubt a grinding wheel and not a lathe cutting tool.

Most of the research and other investigative work in abrasive uses has been carried out on cylindrical grinders. Even when the type of grinder is

Fig. 6-1.   Basic center-type cylindrical grinder, with square-faced wheel at a right angle to the work, which traverses. For grinding tapers, sometimes the wheel is swiveled, sometimes the table. (*Hitchcock Publishing Co.*)

not specified in an abrasive research project, it is most likely to be a cylindrical grinder. A considerable percentage of the literature concerning the causes and correction of grinding problems, as one example, was originally developed for cylindrical grinders and then modified as needed for other types.

Cylindrical grinders being considered here include between-centers plain types (Fig. 6-1), on which the work is held and rotated between centers and usually traversed between the centers, back and forth across the square face of the abrasive wheel or of a belt (Fig. 6-2); angular types (Fig. 6-3), on which the head is usually mounted at something other than a 90° angle to the axis of the piece-part, which may or may not be traversed during grinding; and centerless grinders, on which the piece-part is supported on top of a bar called a work rest between two abrasive wheels. Another variation for external grinding is the superfinishing or microstoning machine, on which the grinding is done by reciprocating stones with concave grinding faces

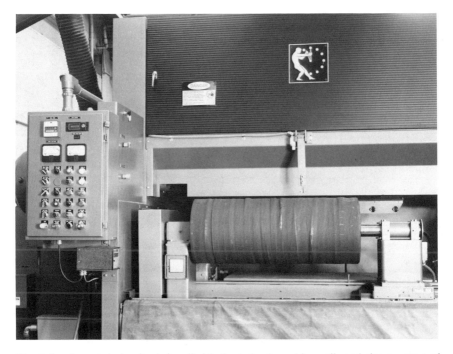

Fig. 6–2.   A coated abrasive belt cylindrical grinder for rubber rolls and sleeves, most of which can be supported on a central shaft. Completely automatic once the roll to be ground is loaded, this machine is reported to be five times faster than the one it replaced. It uses a relatively fine 150-grit silicon carbide belt. (*Timesavers, Inc.*)

shaped to the arc of the round piece-part and held in a framework that presses them inward. (Fig. 6–4.)

There are two types for interior finishing. One is the internal grinder (Fig. 6–5), which finishes holes or bores in piece-parts with a rotating wheel. The other is the honing machine (Fig. 6–6), with rounded sticks of abrasive supported in a holder which presses them outward against the interior wall of, say, an automobile cylinder, the honing motion making them move both up and down and around to finish the cylinder.

CENTER-TYPE CYLINDRICAL GRINDERS

Almost a century has elapsed between the rickety converted lathes mounting 3- × 1 × 16-inch wheels and the recent huge machines with their 24- × 3-inch and even larger wheels. It has taken even longer to convince the in-

Fig. 6–3. This wheel is mounted at an angle to the axis of the work, permitting the easy grinding of a shoulder, or with the correct angle dressed on the wheel face, traverse grinding similar to the machine above. (*Cincinnati Milacron.*)

dustrial world that a grinding machine can be large and at the same time be very precise. Large size and the ability to remove large quantities of metal or other materials seem to be a much more easily accepted concept. As a matter of fact, the stability and the lack of vibration that comes from considerable mass in the machine design promote both stock removal and precision finishes on the same machine and often with the same wheel, which can be dressed "open" for stock removal and "dulled" for finishing. It is reminiscent of an old ad slogan promoting, as I recall it, piston rings: "Tough, but oh, so gentle."

Plain Cylindrical Grinder

The most common type of cylindrical grinder—the plain cylindrical grinder (see Figs. 6–1, 6–2)—comes in sizes ranging from small ones that are used in toolrooms to huge grinders, about half a block long, that are used for grinding steel mill rolls. The wheel spindle is mounted parallel to the axis of the work, which is mounted between centers (cones of metal of appropriate

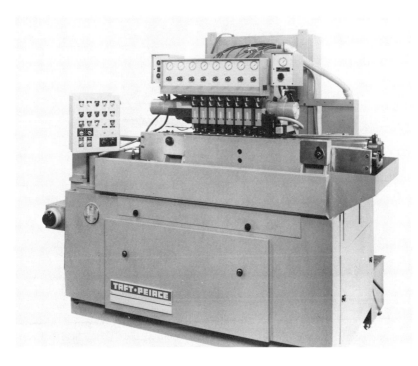

Fig. 6–4. Overall look at a microstoning or superfinishing machine with multiple heads. This process is also done with attachments which fit into, say, a lathe. (*The Taft-Peirce Mfg. Co.*)

sizes) inserted into center holes, like countersunk holes, drilled in either end of the shaftlike piece-part. The centers, in turn, are held in the head (or driving) stock and the tailstock, which support the piece-part (Fig. 6–2). In this setup, as in all cylindrical or centerless grinding, the wheel and the piece-part both rotate. On a center-type grinder, the piece-part obviously rotates between the two centers. Four modes of work-wheel motion are possible. First, the piece-part customarily reciprocates (or in the trade term, traverses) back and forth from left to right and back, while the wheel rotates in a fixed position. Or, second, both the work and the wheel rotate in fixed positions in the mode termed plunge grinding. In basic plunge grinding, the piece-part must be shorter than the wheel is wide. While plunge grinding can be done on any plain cylindrical grinder, it is usually a job for special machines in high production situations, where there is no provision in the machine design for traversing the piece-part. The third method is a combination of the two, in which the work is ground to size in a series of overlapping plunge cuts, usually with one or more final traversing passes to smooth out the small differences between the plunge grinds.

Fig. 6-5. Sizable parts can be internally ground, as in this setup, as the part in front of the machine indicates. Note the size of the wheel, just about as large as the hole will accommodate. (*Hitchcock Publishing Co.*)

This is sometimes called step cylindrical grinding. Finally, when the work is so heavy or so long that traversing it is a major problem, as in the regrinding of large steel mill rolls, the wheelhead is traversed rather than the work. On such machines the operator usually rides along with the wheel, so that he is always in position to observe what is going on at the wheel-work interface.

One point of importance regarding cylindrical grinding as opposed to other forms of grinding is the following: on virtually every other form of grinding, if the downfeed (or infeed) of the abrasive wheel is, say, 0.001 inch, then the reduction in thickness of the piece-part is likely to be the same, or possibly a hair less. But in any cylindrical grinding, if the infeed is the same, 0.001 inch, then the reduction in diameter is *twice* that amount, or 0.002 inch, because stock is removed from both ends of any diameter directly involved.

Surface pressure of the wheel on the work could be high in *surface* grinding, as has been noted, because of the backup strength supplied by the magnetic chuck, even though the crosswise pressure on the work might be limited by the hold of the magnetic chuck on each piece-part. The holding

Fig. 6-6.   Big vertical honing machine, with capacity to hone parts to 15 inches diameter and 60 inches high, needs to be somewhat more than 10 feet high to rough and finish the entire interior surface of the work. Honing is accurate and fast, and produces a useful surface pattern. (*C. Allen Fulmer Co.*)

power of the chuck could of course be increased in some cases by steel blocking devices or in extreme cases by clamping, although clamping does reduce one of the principal advantages of surface grinding—quick setup.

In *cylindrical* grinding, the surface pressure of the wheel on the work is limited by the hold of the centers on the piece-part; if too much pressure is applied, then the piece-part may simply pop out from between the centers. Furthermore, any time that the piece-part is long and thin, that is, when the length is many times the diameter, there is the possibility of its bowing under pressure, thus leaving the geometry of the finished part an unknown quantity. If the diameter of the piece-part is equal to its length, or if length is only a small multiple of the diameter, then there is probably only a small likelihood of trouble; but if the diameter is only a small fraction of the length, then the piece-part must be supported by one or more work rests or steady rests, attachments which fit between the work and the table and keep the work from bending under pressure.

Angular Grinders

The discussion has thus far proceeded as if the principal use of cylindrical grinders were to grind straight shafts. There are obviously a great many such types of piece-parts, but there are also numerous other shapes, tapers for example, and a good many with shoulders, in which the side of the shoulder and a minor diameter of the shaft have to be ground at the same time and usually to a specified radius where the two meet.

The old method of grinding such a piece was to use a straight, type 1 wheel that was traversed along the shaft until the side of the wheel encountered the shoulder, thereby grinding the two at the same time. This caused considerable wear on the leading edge of the wheel, so that by the time it encountered the shoulder, the resulting corner radius was chancy at best. Its size could not be controlled.

This difficulty has been met by the development of the angular cylindrical grinder (see Fig. 6–3), which has a grinding wheel mounted with its axis at an angle of 30, 31, or 45° with the work axis [instead of the 90° of the plain grinder (Figure 6–1)]. With the wheel so mounted and its face dressed to a 90° V, one side grinds the shaft while the other grinds the shoulder, with much better results than those which come from the straight wheel. Of course, grinding the two differing surfaces may tend to produce unequal wear on the two sides of the 90° V face of the wheel, but if this occurs, one remedy that can be used is a "sandwich" wheel in which the half that grinds the shaft is made of slightly finer grit and a slightly harder grade than the half which grinds the shoulder.

This brief discussion by no means exhausts the possibilities of plain and angular O.D. grinders. For instance, on a part that has two separate major diameters that must be ground to match, it may be possible to finish both diameters on a single wheel if they are close enough together, but the more likely setup would have two matched wheels mounted on the same spindle and separated by a spacer of appropriate thickness. The dual-wheel setup, in comparison with a single wheel for both diameters, saves on abrasive wear.

If a formed diamond dresser is used, form, or profile, grinding (the first is the preferred term) is a possibility with a standard plain grinder. Crush form grinding, however, which is a feasible method in cylindrical grinding as it is in surface grinding, is done on specially designed cylindrical grinders because of the pressures involved in dressing the sheels to shape. Form, or profile, grinding will be covered in a later section of this chapter. It is entirely possible that the outlook for grinding shaped piece-parts to final form from the solid is even better and more profitable in cylindrical grinding than it is in flat-surface grinding.

Centerless Grinders

About the only resemblances between O.D. cylindrical grinding and centerless grinding is that both are methods for making round parts and both use abrasive wheels of similar specifications. The grinders are of considerably different designs. Centerless grinding requires two abrasive wheels instead of the one needed in center-type cylindrical; and on the centerless grinder, the work is supported between the two wheels and on top of a work rest rather than between centers (Figs. 6–7, 6–8). For reasons which will be explained later, centerless wheels are thicker than are most other wheels; they may, indeed, be the thickest wheels made (Fig. 6–9).

Centerless grinding is a comparatively recent development, dating, in terms of industrial acceptance, from the early 1900s. The idea may well go back a century or so, but general use in industry started about 1915 or perhaps 1920. In fact, centerless grinding has been primarily a replacement for cylindrical grinding, principally because it is a much faster method for machining piece-parts on which it can be used. Here are some of the reasons advanced by proponents of its use:

Fig. 6–7. A somewhat different view of a throughfeed centerless job. Though centerless grinding can handle parts with shoulders, tapered parts, and other like these, its principal use is that of a throughfeed process. (*Hitchcock Publishing Co.*)

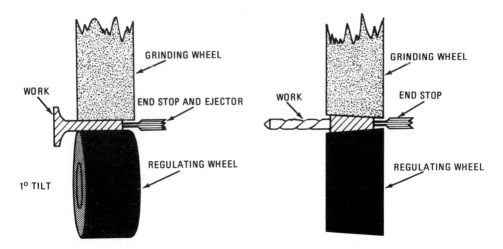

Fig. 6–8.   Two other types of centerless work. On the left is a part resembling a valve stem, which is ground to the shoulder and then reversed to eject from the same side it entered. The other is a drill shank, which handles essentially the same way. Note the angled grinding faces of the wheels on the right. (*Bay State Abrasives, Dresser Industries.*)

1. In through-feed operation (on parts like piston pins, which are straight cylinders without shoulders or other protrusions) grinding is continuous. Loading or unloading can be done during grinding and does not, in fact, have to be done manually.
2. Each piece-part is rigidly supported along its entire length. This permits heavier pressure and, as a consequence, faster stock removal.
3. There is no danger of springing or bending long, slender piece-parts. Indeed, part length is not a factor. Quite some years back, when I asked the owner of a centerless job shop what was the longest part he could grind, he grinned and replied, "How far is it to the street?" For long lengths, of course, outboard supports to hold up the ground section are desirable.
4. Roundness is generated from the diameter rather than from the center, so there are no problems involved in centering, and less stock needs to be left for grinding. This saves time, material, and abrasive.
5. Errors in setup and in adjustments for wheel wear are cut in half, because stock removal is measured on the diameter rather than on the radius.
6. Errors due to wheel wear are less serious.
7. A centerless grinder is a comparatively simple machine, with few wearing parts. This makes for low upkeep and long machine life.

Fig. 6-9. The sheer size of centerless wheels is something to catch the eye. When this grinding wheel is lowered into place by an overhead crane, the top guard will be swung down into place and the grinder will be ready to roll. (*Cincinnati Milacron.*)

How the Grinder Works.

The action of any machine tool is usually pretty straightforward, and not particularly of broad interest. The centerless grinding action, however, has no counterpart anywhere else in machining, so a paragraph or two of description can be of interest to the reader.

As was said earlier, the piece-part(s) are held between two abrasive wheels on top of a work rest (or work blade). One of the wheels, the grinding wheel, is usually vitrified-bonded and operates at standard speeds—6500 sfpm to perhaps 8000 or 8500 sfpm. The other, or regulating, wheel, usually rubber-bonded even for abrasive belt centerless grinders, runs at much slower speeds and prevents the work from spinning, as it would if both wheels ran at the same speed. Single-diameter parts like piston pins and bar stock are through-fed, entering at one side of the wheels and exiting at the other. The grinding of bar stock, which comes in long lengths, was, incidentally, the first wide-spread application of the centerless grinder. It provided this industry with a better product that was more uniform in size, while removing the decarburized skin and exposing the seams which can develop in the rolling process. These examples are from among the many relatively small diameter parts that are being machined at high-production rates by through-feed centerless grinding. The progress of the piece-part along the work rest and between the wheels is controlled by tilting the axis of the regulating wheel and adjusting its speed.

But don't conclude that only this kind of small, single-diameter work can be handled. Solid bars up to about 4 inches in diameter can be ground, as can much larger diameter parts if they are not too long. It is the weight of the piece-part that is the controlling factor, not its nominal size; and if centerless grinding is a desirable process because of other factors, it is frequently possible to rig a combination to permit the parts to be accommodated.

Cylindrical parts with shoulders, and tapered or formed parts (Fig. 6–8), can also be centerless ground, using the infeed technique. In this approach, the grinding wheel and its regulating wheel are match-dressed to the taper or form of the part, the regulating wheel is skewed just a little bit to hold the part against a stop positioned to locate the part in relation to the wheels. Parts may be fed from a magazine, or they may be placed between the wheels by hand. One of the wheels, of course, has to be backed off a little before this can be done. The stop may also be used for ejecting the part after it is ground.

Out-of-balance work can also be machined efficiently on centerless grinders, because no time need be spent on finding the centers. The part is rounded up from its own diameter rather than from its center. The work material is no problem. All kinds of metals and most nonmetallic materials—glass, hard rubber, plastics, rawhide, and porcelain, to mention some of the common materials—can be centerless ground. Precision glass tubing is a very common example. The secret of grinding any material lies much more in the selection of the grinding wheel, and possibly of the coolant, than it does in the capabilities of the process itself. In fact, in any type of grinding, the question of whether or not a material can be ground lies much more in the development of an adequate wheel specification than it does in any other factor.

One element that is peculiar to centerless grinding is that while piece-parts are being machined, they move across the face of the grinding wheel. This gives the thickness of the wheel more importance than it has in other methods of grinding, and it has led to the development of wheels 20 inches or more in thickness, certainly much thicker in relation to wheel diameter than any other type of wheel. The thicker the wheel, of course, the greater the potential grinding period. The thickest wheels are made by cementing together two halves that are cut on the bias to avoid objectionable lines on the surface of the finished part. Lines could result if two flat discs of abrasive were simply cemented together.

Abrasive belt centerless grinding (Fig. 6-10), however, can compete successfully with wheel centerless installations, and of course with some turning operations, as is illustrated by the following example.

The part in question is a bushing made up of an inner steel tube and an outer steel tube separated by a rubber cushion. During the assembly of the bushing, high compression of the rubber could cause bulging of the outer tube, an irregularity that the wheel centerless grinder would not accept until the part(s) went through an added sizing operation which required disassembly of the bushing. But since the abrasive belt grinder would accept parts with bulges of as much as 0.040 inch and still remove the irregularities and bring the finished part into the ±0.001 inch required, the extra sizing work was completely eliminated, a considerable saving.

As a bonus on this installation, some similar parts that had been previously turned on a turret lathe were also routed to the belt centerless grinder.

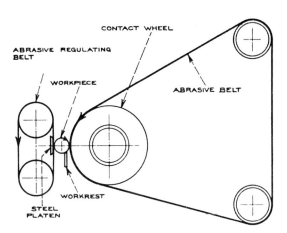

CONTACT WHEEL

ABRASIVE REGULATING BELT

WORKPIECE

ABRASIVE BELT

WORKREST

STEEL PLATEN

Diagram shows relative position of work piece, feed belt, work rest, and contact wheel.

Fig. 6-10. A belt centerless grinder operates much like a wheel machine, with the contact wheel replacing the grinding wheel, and, of course, the backstand or idler pulleys. There may be a standard regulating wheel, or there may be a sort of platen grinder backup with another and smaller belt. (*White-Sundstrand Machine Tool Inc.*)

Superfinishing

Superfinishing is a method of refining the external surface of a cylindrical part by reciprocating an abrasive stick shaped to the approximate arc desired in the finished part back and forth along the length of the part as it rotates. The action is similar to honing, an internal cylindrical process which will be discussed shortly. The stick, which is under mild pressure, can produce a low microinch finish, almost on a par with that produced by centerless lapping. It removes scratches, smears, grinding ridges, and waviness resulting from preceding operations; it improves part roundness, and improves any bearing surface. The process has been improved to a point where it can be applied directly after finish turning, making it a process for stock removal as well as fine finishing. (See Fig. 6–4.)

Internal Grinding

Machining the inside diameters of parts has several obvious differences from O.D. machining. First, an internal grinding wheel must naturally be smaller in diameter than the surface it is finishing (Fig. 6–11). Second, the area of abrasive contact in internal grinding is greater than it is in external grinding. External grinding is a line contact between two outside diameters. Internal grinding matches an external diameter on the wheel against the internal diameter of the piece-part, causing an area contact of much greater size. Third, internal grinding wheels must operate at a much higher rpm than external wheels to achieve efficient surface speed on the grinding surface of the wheel. Some of the smallest wheels, sometimes termed mounted points, may rotate at speeds in excess of 100,000 rpm. Fourth, the wheel has cantilever support whether or not it is mounted on a mandrel; the nature of internal grinding makes this necessary because the wheel must always enter the piece-part from the open end, and the wheel can never be supported on both sides. The other end of the hole is always blocked whether it is a blind hole to begin with or whether the blockage is the face plate against which the piece-part is mounted. In fact, the work surface entirely encloses the area in which the grinding must be done. And if coolant is indicated, it must of course also be applied within this area, which is not always an easy thing to do. Thus, internal grinding is a unique form of machining with abrasives, a process that has its own unique set of problems.

It is entirely in order to consider why a process with these limitations should be considered at all. One major reason may be that in many situations, either because of the required precision of the fit between mating parts or because of the hardness of the work material, there is really no

Fig. 6–11. The small size of the wheel points up one of the problems of internal grinding: the spindle must rotate at a tremendous rate in rpm's for the wheel face to approach an efficient grinding speed. On a wheel of this size the rate might be as much as 60,000 to 70,000 rpm. (*Hitchcock Publishing Co.*)

other choice. Internal abrasive machining, either by a rotating abrasive wheel or by the combination reciprocation and rotation of honing, is consistently the most accurate method of finishing holes where a metal-to-metal seal is required or where close-fitting assembly of parts follows the machining.

Internal grinding has these additional advantages and limitations. Surface finishes, generally speaking, can be as good as they need to be. Cutting force is low, making it possible to machine thin-walled delicate parts without severe distortion. Profiles or internal forms can be produced efficiently, often with relatively inexpensive truing devices. And as with all other abrasive machining processes, accuracy in form, dimensions, and locations is consistently better than with any alternate machining method. And, of course, abrasives will machine hard materials that no other method can touch or handle with efficiency. Internal grinding can readily be automated with rapid loading and unloading devices, continuous in-process size checking, and fast-acting workholding and locating devices.

As a matter of efficiency, however, it is advisable to have the work as close to finished size as possible when it comes to the internal grinding operation. Internal grinding is not outstanding, except perhaps in a relative sense, as a means of removing stock. However, finding the optimum is, of course, a matter of balancing the relative costs of rough machining or forming to close tolerances against the cost of grinding off the excess stock. No machining operation with abrasives *requires* significant excess stock for reasons of process capability.

It should be obvious also that internal grinding will not be really efficient if, for example, there are substantial variations in the stock to be removed from one part to another. This is true of all abrasive machining operations, but probably because of the size of the abrasive tools involved, it is more significant in internal abrasive operations than in external. Much the same can be said about other poor piece-part conditions—inaccurate locating surfaces, out-of-round holes or distortion—caused by lack of control in a previous operation like rough machining or heat treating. There are definite limits, again, to which any abrasive machining operation can compensate for deficiencies in preceding operations (and as the abrasive operation is frequently the final one, it often gets the blame for those earlier deficiencies), but internal grinding and honing have perhaps less leeway than does outside diameter or flat-surface grinding.

It is probably safe to say that most of the troubles in internal grinding result from the fact that it involves a relatively small wheel on the end of a spindle (or in the case of mounted wheels and points, on the end of a mandrel) revolving at high rates of speed. These conditions tend to intensify the negative effects of any runout or vibrations in either the workhead spindle or the grinding wheel spindle.

Most of the jobs done by internal grinders are round, uninterrupted, plain holes whose axes are also the centers of the whole piece-parts and whose walls, consequently, are uniform in thickness. This is work that might well be classified as metallic rings of differing dimensions. And for grinding, the piece-part is centered on a backing plate and rotates while it is being machined. However, this is not always the case. The hole may be off-center; there may be slots across the path of the grinding wheel; and the part may not rotate at all. In fact, it is even possible to do a limited amount of external cylindrical grinding where the surface is otherwise not accessible for machining. It is not a particularly effective method, necessarily, but it may be the only feasible one. It has been said many times, of course, that abrasives will do anything that cutting tools will do except drill a hole, but once the hole is there, internal grinding of some type will almost always finish it to close tolerances and required surface quality.

Honing

Strictly speaking, the term honing, in industrial use at least, ought to be limited to describing a controlled, low-velocity abrasive machining process for enlarging a hole to precise dimensions. But, like lapping, honing has long been used as a convenient term for a number of fine finishing operations. However, external lapping is, in principle at least, a superfinishing operation; and flat honing is actually lapping, if the abrasive grain is loose, or a very fine level of disc grinding if the grain is bonded in a flat disc. Honing will be considered here as an internal abrasive machining method.

Velocity is the principal difference between internal grinding and honing, and the differences in velocity have made differences in the kinds of applications each is suited for. In internal grinding, as was explained earlier, the surface speed of the abrasive wheel is at least 6500 sfpm, and may be higher. (The fact that the piece-part itself acts as a safety guard around the wheel improves the safety of the operation and is generally considered justification for somewhat higher wheel speeds than are generally used. The honing stones, operating in similar workpieces, move at a slow 80 to 300 sfpm.).

A honing tool consists of a core or body, a cone, and a set of honing stones in holders, which are inserted in slots around the perimeter of the body (Fig. 6–12). In operation, the cone acts as a wedge in the center of the tool, forcing the stones out radially with equal pressure on each stone.

The tool is attached to a spindle of a mechanism (Fig. 6–13) that imparts to it a rotating/reciprocating motion against the surface being honed. This is a unique motion in machining; it produces in automotive cylinders (where honing is virtually the universal final machining process) a surface pattern of interlocking minute scratches that retain a film of oil on the surface, which is, in turn, the principal reason why new cars have for quite a few years been able to operate at cruising speeds right from the first mile without having to go through a break-in period as was formerly necessary. (The break-in period primarily gave the cylinder and its piston an opportunity to wear into a fit with each other by smoothing out the ridges left on the mating surfaces by the machining methods then available.)

The environment in which honing is done dictates that there be a few variations from so-called standard abrasive machining. For example, because the honing stones cannot be effectively dressed or resharpened once they are in the holders, they must be soft enough in grade to be self-sharpening, even though this leads to extra abrasive wear. Moreover, the stones are cut essentially to the arc of the cylinders on which they will be used, so there will be virtually full contact from the beginning between the

Fig. 6–12. This is a honing head, which is designed so that the bonded abrasive sticks are forced out against the walls of the hole or cylinder that is being honed. As the head is rotated and reciprocated simultaneously, the sticks work first on the high spots and, as these are blended in, gradually increase the work area until the whole inner surface is being honed. (*Hitchcock Publishing Co.*)

cutting surface of the honing stones and the surface of the cylinder. Because of this and because of the continuous resharpening of the stone in use, stock removal by honing is high even with the low velocity of the honing tool.

To ensure that the cutting pressure is equalized on all parts of the surface, one of the elements—either the piece-part or the tool—must float to provide a self-aligning action. The floating action also eliminates the need for perfect alignment and the possibility of a shift in location of the bore.

Piece-parts coming to final honing usually have defects stemming from heat treating, forming, or previous machining operations. They may be tapered, out-of-round, bell-mouthed (bigger in diameter at the ends than in the middle) or barrel-shaped (which is the opposite), or wavy. There is obviously a limit beyond which such defects cannot be corrected; but for piece-parts within acceptable limits, the length of the stones and the rigidity with which they are held in the tool result in their cutting only on the high spots until correction has been made and the full surface has reached the same diameter.

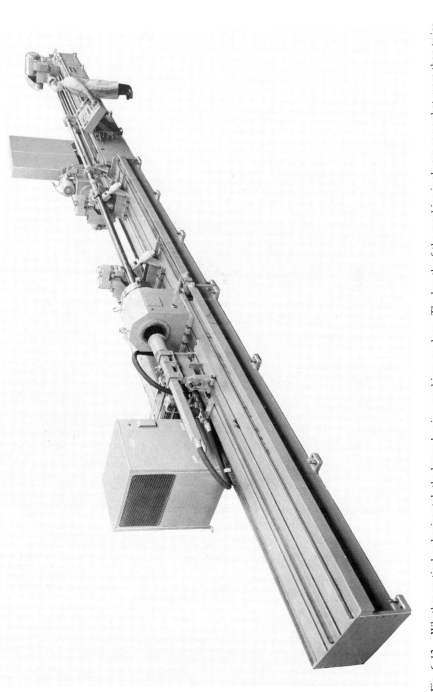

Fig. 6–13. Whether vertical or horizontal, the larger honing machines are long. The length of the machine is always somewhat more than twice the longest piece-part it can handle, and this one removes about 0.100 inch stock from the inside of tubing. (*Hitchcock Publishing Co.*)

Internal cylindrical surfaces from 0.125 inch to 42 inches in diameter and lengths from 0.125 inch to 50 feet long are being honed, including parts with blind ends, keyways, or tandem bores. On the very long bores, the stock-removal ability of honing exceeds that of all other processes.

Superfinishing or Microstoning

Superfinishing and microstoning are in principle the same operation. Both terms refer to the finishing of an external cylindrical surface that is rotating under pressure from a shaped abrasive stone (or of several such stones) and reciprocating as in honing. In other words, it is a sort of mirror image of internal honing. As in most precision grinding, the stones are vitrified-bonded, usually by either aluminum oxide or silicon carbide, depending on the piece-part material.

The process—superfinishing or microstoning—dates from about the time of World War II. The method has achieved the capability of improving finish and correcting geometric errors resulting from previous machining—chatter marks, feed spirals, and out-of-round conditions, for example—directly from the lathe and without intermediate grinding. Figure 6–14 shows how a microstoning attachment is mounted on a lathe, a common use.

The principle is simple. Stones shaped roughly to the arc of the piece-part to be finished are oscillated against the piece-part either on a special finishing machine or on a planer, lathe, milling machine, or boring mill equipped with a finishing attachment. The abrasive stones are constantly being worn away, so that each of them soon follows the exact contour of the part in process. This wraparound action tends to correct lobing or any other out-of-round conditions. And because it is the roughness of the work that wears away the stone, there is little reduction of base metal once smoothness has been achieved.

Microstoning equipment, one of the major types available, is pneumatically operated. The stones are held against the work by air pressure; air pressure also raises and lowers the stones (there may be more than one working on a piece-part at one time), so that there is very little vibration to distort the final finish.

Crush-Form Grinding

Crush forming is actually a method of dressing abrasive wheels, which is a factor in surface grinding as well as in cylindrical, centerless, and thread grinding but given special emphasis with respect to center-type cylindrical grinding because specially designed machines have been developed for this

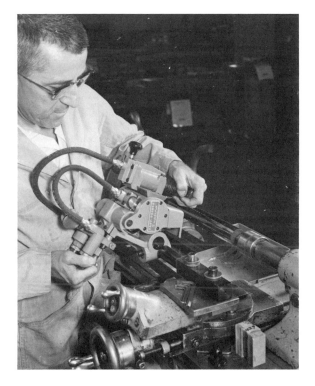

Fig. 6–14. Microstoning, sometimes called superfinishing, need not be an expensive process. This microstoning attachment has been designed to fit on a lathe. The operator has the head in his left hand. (*Taft-Peirce Mfg. Co.*)

type of grinding. In this type of application, crush forming has had its greatest effect.

In essence, crush-form cylindrical grinding is very simple. Two or more rolls duplicating the form desired in the finished parts are made. One of these is designated the master roll, and the other(s), work roll(s). At the start of a production run, the operator manipulates the grinding wheel over the work roll so that the reverse of its form is crushed into the grinding surface of the wheel. When the wheel is properly form-dressed, it is adjusted to grind piece-parts (Fig. 6–15). When the form in the wheel is out of tolerance, the wheel is re-dressed on the master roll. Next the wheel is used to regrind the work roll to form. This general sequence is repeated throughout the production run. It is a most productive and efficient system, because parts can frequently be ground directly from blanks, or the "solid," in one operation (Fig. 6–16).

Crush-form grinding does have some rather definite requirements,

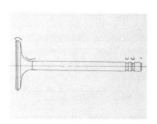
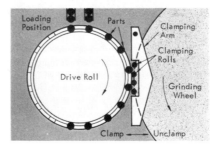

Fig. 6–15. This is essentially a crush grinding job with a twist. These automotive valve stems are rotated against the formed wheel by the drive roll. (*Bendix Corp.*)

however. For one, the grinder, of whatever type, must be rigid enough to withstand the substantial pressures of crushing. For another, the wheel must be kept free of chips and grit. (One of the major suppliers of this type of machine thinks that this requirement is important enough to have designed into the back of the machine, away from the work area, a high-pressure nozzle directed at the wheel to remove any possible loading and keep the wheel grinding face clean with a spray of coolant.) And crush forming is almost exclusively used with vitrified-bonded wheels.

Fig. 6–16. The relationship of the piece-part in front of the wheel and the crush roll above stands out clearly. The grooves in the wheel are thin, and the lands are wider. (*Bendix Corp.*)

The crush-dressing action, as is easy to visualize, leaves many jagged and sharp points and edges of abrasive which cut fast but do not provide the smoothest of finishes. Finish can readily be what many would call commercial, however, and thus sufficient for most of the uses of the resulting parts. For most operations where crush-form grinding is an attractive possibility, the ability of the wheel to cut fast and efficiently and accurately outweighs any lack of fine finish.

Another advantage of crush forming is that it makes possible the abrasive machining of some jobs where diamond dressing is not possible (Fig. 6–17). Furthermore, it is a relatively cool machining job, because the fractured points of crushed abrasive cut fast and cool. It tends to be

Fig. 6–17.   On this job, a steel mill roll, the grooves and the lands are about equal. Crush form grinding is quite adaptable. (*Bendix Corp.*)

economical of material, because the piece-parts can be brought to the grinder with the minimum of extra stock of which prior machining operations are capable. And a crush-dressed wheel takes less power than does a diamond-dressed wheel.

The principal disadvantage of crush dressing is its inability to hold a very sharp edge, either in the valley or on the crest of the form. A radius of about 0.002 or 0.003 inch is about the best that can be achieved. Grit size is in the wheel ranges from 120 to 220, depending on the smallest radius that must be machined. And, finally, grade is medium, K to N.

With a grinder that is built to take the pressure, crush-form cylindrical grinding is a very efficient and economical procedure, particularly when the form can be ground from the solid, without intervening machining operations. The possibilities are virtually unlimited. Some are illustrated as in Figure 6–18; others need only a few words of description. For example, one

Fig. 6–18. An informative collection of crush ground parts. (*Bendix Corp.*)

operation that can be done by crush-form grinding involves the simultaneous grinding of 49 diameters on a shaft; it's rather doubtful that there is an alternate method—and certainly none in the same price range. Finish is not in the honing or lapping range, but it's adequate for most purposes. Crush-form grinding is worth considering when almost any form within its capabilities is being looked at.

Universal Cylindrical Grinders

Whereas the earliest grinders used in industry, as has been suggested a number of times, were probably plain cylindrical grinders converted from lathes, it is probable that most of the grinders in use in the nineteenth century were universal types on which both external and internal cylindrical grinding could be handled, and, in a pinch, surface grinding of a sort. These were not production machines; they were for use in the toolroom only. They were not even expected to be economical, but they had to turn out precision work.

The term "universal" is used today for practically any grinder that will do a little more than one limited kind of grinding, but generally it is used to describe a toolroom grinder on which even as small a number as half a dozen would look like a long production run.

SUMMARY

The variation in types of grinding machines to produce cylindrical parts is not quite as large as those for making flat surfaces, but it is impressive. Considering everything, it seems logical to conclude that the largest gains are going to come in abrasive belt centerless grinding and in crush-form wheel grinding. Effectiveness in centerless grinding depends to a considerable degree on the length of time the piece-part is exposed to the abrasive. For wheels, the limit in the length of this abrasive "path" is about 2 feet, which has been achieved in a grinder designed so that the wheel is supported on both ends. This length is not likely to be increased by very much. Of course, there is the possibility of tandem lineup of grinders, but this method requires a considerable quantity of production for its justification.

On an abrasive belt centerless grinder, on the other hand, there are no comparable problems of weight or manufacturing difficulty in lengthening the path. Belt centerless grinders in tandem are relatively easy to put in place, more so than tandem wheel grinders.

As for crush-form wheel grinding, the number of multiple-diameter shafts—and for that matter of shaped but essentially flat surfaces to which crush-form grinding is also applicable—is increasing yearly. Use of this method should increase at least as fast—quite possibly faster—as the increase in potential applications, as the efficiency of crush-form grinding is more widely recognized.

7
Specialized Grinding Machines

Most abrasive-using machines, like, for that matter, most machine tools, can be divided into those that produce flat surfaces, the subject of Chapter 5, and those that produce essentially cylindrical surfaces, discussed in Chapter 6. However, there are a number of grinding machines which do not fit readily into either group but which ought to be described. There are in addition some machines so specialized that they are of concern only to narrow segments of American industry.

The grinders included in this chapter include abrasive wheel cutoff machines, electrochemical grinders, portable grinders and coated abrasive grinders which use some other form of coated abrasives than belts. Machines using loose abrasive grain as the cutting medium might also have been included here, but they are so different in so many respects that it seemed logical to devote a separate chapter to them.

ABRASIVE CUTOFF MACHINES

The use of thin abrasive wheels to part materials has been a practice for many years, and for good reason: abrasives could cut materials that no other cutting method could touch. The thinnest cutoff wheels—often just a few thousandths thick—are rubber bonded. Thicker wheels, up to 3/8-inch, are resinoid-bonded. Both bonds can stand the very high speeds which are the norm in this operation. Abrasives are usually aluminum oxide and silicon carbide, with some thin diamond wheels for carbides and

CBN wheels for some of the steel alloys. Some applications are shown in Figs. 7–1, 7–2, 7–3 and 7–4).

Comparisons

Compared to other material-parting methods, abrasive cutting has several advantages:

1. It has the ability to cut much harder materials than any other method.
2. It produces two finished sides per cut. The others leave burrs.
3. Abrasive cutting is fast; in fact, it probably becomes relatively faster as the material to be cut gets harder, which is just the reverse of what happens when competing methods are used.
4. Abrasive cutting is not likely to harden the work.
5. On a flat surface, the abrasive wheel can cut very close to the adjoining surface, leaving only a little metal to be removed later.

Fig. 7–1.   Rough cutoff of excess metal from a casting. This wheel rotates safely at an approved speed of probably 16,000 sfpm. (*ACCO Industries, Inc.*)

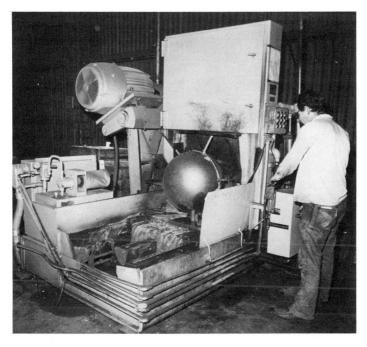

Fig. 7-2. Automatic cutoff machine for cutting off the risers (projections) from the crusher ball. (*ACCO Industries, Inc.*)

The principal limitation of the abrasive cutoff machine and wheel is probably that the wheel must cut in a straight line, as there is no way to make it cut around curves or corners.

In connection with the other principal methods of cutting—shearing, flame cutting, and band sawing or cold sawing—the following observations are in order.

If the quality of the cut made by a shear is satisfactory, then there is no competition. Under such conditions, shearing is the best method, being the cheapest.

Flame cutting is fast, but it leaves a wide kerf behind, thereby wasting material, and it does not sit well with environmental regulations. Furthermore, when the cost of the gas is included, as it should be, flame cutting is not always competitive with abrasive cutting. On castings it is essential to keep the flame well back from the body of the casting to reduce the risk of undercutting, a practice which may leave excess metal to be removed in snagging.

In comparison to band sawing or cold sawing, abrasive cutting is faster and leaves a smoother surface on the sides of the cut. As the metal gets

Fig. 7-3.   Swing-frame (for obvious reasons) cutoff machine. This is a very maneuverable machine for rough cutting. (*ACCO Industries, Inc.*)

harder, the difficulties of the saw increase. But band saws can cut contours.

Thus, if only random or casual cutting is involved, then the choice might well depend on the relative volume of hard materials to be cut and the nature of the cuts. However, abrasive cutting is not limited to just this sort of cutting; it has distinct possibilities as a production cutting operation. For example, if you need quantities of metal slugs cut from bar stock, it is virtually no problem to convert a simple manually operated chop-stroke abrasive cutoff machine to entirely automatic operation, with an employee needed only to monitor the operation at intervals and insert a new length of bar stock when the old one is used up. Almost any machine shop or maintenance shop can make the conversion. Of course there are machines built for this use, which is fairly common. And the cuts will be made, to reiterate, with a minimum loss of stock and with two finished sides. If burrs occasionally occur, or if the slugs must be entirely burr-free, a run through an automatic vibrator that uses loose abrasive grain will easily remove them—and without significant labor cost.

Fig. 7-4. Diamond saws are very precise cutting tools—for very hard, mostly nonferrous materials. Note that the diamond is simply a thin coat on the outer rim of a thin disc. (*Engis Corp.*)

Types of Cutoff Machines

There is possibly a question about whether most abrasive cutoff machines qualify, in most people's minds, as machine tools. In the most popular type, the manual chopstroke unit, all that is needed are a pivoting arm on which the wheelhead and the wheel and its safety guard are mounted, a power source to drive the wheel, a clamping device to hold the piece-parts, and a frame to hold the unit together. Many of the smaller machines are bench-mounted. This type is usually operated dry, without coolant, and the size limit for the cut is about 3 inches.

Oscillating types (Fig. 7-5), which add a horizontal reciprocating movement to the vertical downfeed of the chopstroke machine, increase the cutting capacity to 12 or perhaps 14 inches. In wet cutting, oscillation also permits the coolant to enter the cut, thus increasing its effectiveness.

Horizontal cutting machines are built to make long cuts through flat soft

Fig. 7-5.   Horizontal oscillation, shown here in a schematic view, increases the cutting range of a cutoff machine and, if the job is being done with coolant, aids in getting the fluid into the cut where it is needed. (*Hitchcock Publishing Co.*)

or hardened stock at production rates. They are often used for slotting. The wheel reciprocates automatically in moving across the piece-part, and feeds downward at the end of each pass. Slabs on the order of 6 inches in thickness and 30 feet long can be cut routinely, and deeper cuts can be made, if necessary, without any parting lines. Wheels may or may not oscillate, in addition to their other movements.

Rotary cutting is a means of coping with cuts requiring more than the 10-, 12-, or 14-inch capacity of other types of abrasive cutting machines. On the latter three types, the piece-part rotates, but there are machines designed so that the abrasive cutting wheel rotates about the piece-part, which is useful when the work to be cut is quite large and heavy. This last variation is called planetary cutting. Smaller wheels can be used, since the wheel needs to cut only a little more than halfway through the work to complete the cut. This is especially useful in severing pipe or tubing, since the wheel need cut only through the wall.

Probably the ultimate in wet cutting is submerged cutting, with the piece-part completely covered by the coolant. This method is used, for instance, in making metallographic specimens of highly heat-sensitive materials.

Summary

Abrasive cutting or sawing is an efficient means of parting all sorts of materials, particularly the hard ones. And with higher horsepower machines that can generate higher wheel speeds and rates of cut, production is improved, labor costs are lowered, and quality of cut is usually improved. Quality of cut is dependent primarily on the coarseness of the grit in the wheel, and efficiency usually dictates the use of the coarsest grit wheel that will produce a satisfactory finish on the sides of the cut. Reducing cutoff costs to a minimum requires as rapid a cut as the power on the machine will allow. It should be noted that these conditions also reduce wheel life, but then the costs of the abrasive, here as elsewhere, are practically always offset by other savings.

ELECTROCHEMICAL OR ELECTROLYTIC GRINDING (ECG)

Electrochemical grinding is difficult to classify because it is distinguished from other types of grinding by the *method* of material removal rather than by the shape of the workpiece. Its greatest use is probably in the grinding of carbides, an operation usually done by diamond wheels in conventional grinding, and, as it happens, also done with diamond wheels in ECG, but with a lot less wheel wear in the latter approach.

The easiest way to think of ECG is as a reverse electroplating process (Fig. 7-6). That is, material is removed from the piece-part rather than deposited on it. Approximately 90 percent of the stock is removed by electrolytic action; the other 10 percent is ground off, though the wheel action is actually more a matter of cleaning up the surface being worked on rather than of removing any significant stock.

In electrochemical grinding, the workpiece is the anode from which particles of material are removed and attracted toward the cathode, an electrically conductive grinding wheel which is the negative terminal. The piece-part and the wheel help make up a dc electric circuit which is completed by the electrolyte that covers both. The abrasive particles have another role; they help maintain the gap between the conductive portion of the grinding wheel and the piece-part. This gap, which may range from 0.0005 inch to 0.001 inch, is enough to allow the electrolyte to pass between the wheel and

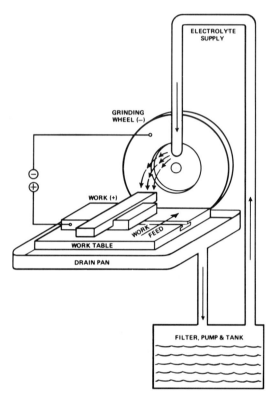

ECG

Fig. 7-6. Illustrates the basics of electrochemical (sometimes called electrolytic) grinding. (*Hammond Machinery.*)

the piece-part, thus ensuring continuous electrochemical oxidation. (Fig. 7-7 may help explain the action.)

It is worth mentioning that the abrasive particles must be nonconductive. If they were not, they would cause short circuits when they touched the piece-part. For this reason, aluminum oxide and diamond wheels are used in ECG applications; silicon carbide wheels are not.

Electrochemical grinding, then, has these advantages:

1. Work is machined without leaving burrs and surface cracks.
2. Because the grinding does not cut the base metal of the piece-part, very fragile and thin parts can be machined accurately.

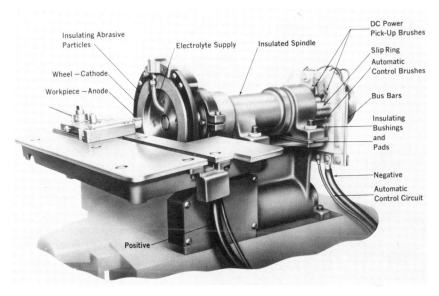

Fig. 7-7. A labeled electrolytic grinder. (*Hammond Machinery.*)

3. Exotic metals which are difficult to cut or machine by any other means can be easily ground by ECG.
4. ECG generates no heat; hence there is no hardening, burning or distortion of the work.
5. Surface finishes of 8 to 20 microinches are easily attained. (In conventional grinding, it is generally considered that a surface finish of 16 to 20 microinches can be achieved without resorting to fine finishing operations or without taking the extra pains and care that increase costs.)
6. Because the basic abrasive wheel function is cleaning rather than stock removal, wheel wear is at a minimum, and wheels last a long time.
7. Forms are readily generated in hard materials.

But with such advantages, ECG has been widely used primarily in aerospace industries and not as much elsewhere as one might expect. Of course, aerospace industries use a greater percentage of hard-to-machine materials than do probably any other industry, and their demand for perfection is preeminent, enough to outweigh many of the cost-effectiveness restrictions that many other industries face. There is still the question of why a technique with so many advantages is not used more widely in general industrial plants. There are probably several reasons for this, among which are the following:

1. An electrochemical grinder is much more expensive, on the order of five or six or even more times as expensive as a comparable conventional grinder. The electric connections, the fact that the whole machine, must be made of corrosion-resistant metals, and similar factors tend to raise the ECG price. Salt solution, the usual electrolyte, is very corrosive, requiring special steels in the machine.
2. Probably many plants simply do not have the stringent quality requirements of aerospace, so their needs can be adequately met by conventional grinding.
3. An electrochemical grinder cannot, without considerable precautions, be shut down for an extended period of time, so, before purchase, it is a good idea to be sure of having enough work for the machine for a considerable time. If the machine is shut down for any length of time, the electrolyte tends to attack and corrode the tank in which it is held.
4. It is worth noting that there are relatively few standard machines currently; many of those which have been manufactured are custom-built for a particular operation. However, with increased use there are certain to be more standard machines available.
5. While it is almost certain that any ECG machine can be used also for conventional grinding, if such a purchase is being considered, it is a good idea to be certain that this can be done.

SUMMARY

Electrochemical grinding could very well be the wave of the future in machining; it has been considered as such for some time. It compares so favorably with milling or grinding on so many counts that wider use almost seems inevitable. But it is also true that the machines are quite high-priced in comparison with the competition, and it is not good management to keep idle a machine that costs in the upper-five-figure range, to say nothing about what downtime of any length does to the machine, as noted earlier. Finally, even though the corrosion that is possible with the electrolyte is probably not a deciding factor one way or the other, it can hardly be considered an unqualified plus factor.

JIG GRINDING MACHINES

Perhaps the least technical way to describe a jig grinder is to say that it is a very sophisticated drill press equipped with a mounted grinding wheel, but such a comparison does little more than suggest the appearance of the jig grinder. Both have vertical spindles operating above a work table; the major difference is that the drill press makes holes but does not particularly

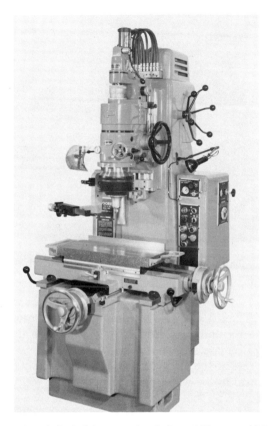

Fig. 7-8.   A modern jig grinder is light years ahead of any drill press, which makes holes. The jig grinder works after the hole has been drilled. (*Moore Special Tool Co., Inc.*)

improve them once they are made, whereas the jig grinder (Fig. 7-8) does not make holes but does a superior job of refining holes already made—irregular as well as round—to very close tolerances. In fact, the jig grinder's best area is probably the kinds of holes needed in die making (Figure 7-9).

Jig grinding demands a combination of three different motions: the high-speed rotation of the grinding wheel held in the grinding head; a vertical reciprocating movement of the whole head of the machine, the range being determined by the length of the hole to be ground; and a slow, planetary rotation of the grinding head axis around the main spindle axis (Fig. 7-10). And on most modern machines, the complete head can also be tilted a few degrees off the vertical to grind forms or chamfers and to "break" corners.

Because the wheels used may be less than 1/4 inch in diameter and mounted on a mandrel the resulting necessity for high rotating wheel

Fig. 7-9.   Closeup of the working area of a jig grinder. All the other mechanism of the machine tool is to ensure that everything goes well here. (*Moore Special Tool Co. Inc.*)

speed—perhaps as much as 175,000 rpm—can cause problems with the mandrel, with the possibility of trouble increasing with the length of the mandrel and any decrease in mandrel diameter.

Jig grinding is one of the most successful and probably the earliest uses of numerical control in the whole field of machining with abrasives.

PORTABLE GRINDERS

Industrial portable grinders (Figs. 7–11 and 7–12), which range from a few ounces to a few pounds in weight, are hand tools equipped with grinding wheels, or possibly belts or other forms of coated abrasives. For rough grinding, they are useful for removing fins, parting lines, and casual defects from castings, and they are suitable for spot removal of defects from big enough piece-parts such that it is easier to bring the tool to the part rather than the reverse. They can use straight, cup, or mounted wheels as well as brushes and even cloth buffs. Equipped with depressed-center wheels, type 27 or 28, large portable grinders are useful for rough cutoff work.

On the other end of the scale, the fine-grit wheels of the small "pencil" grinders are used for fine touchup work on dies.

However, the rough work is so very rough and the fine work so very fine, that it's something like comparing a meat cleaver with a scalpel.

Foundry grinders come in either straight (Figure 7–11) or vertical (Figure

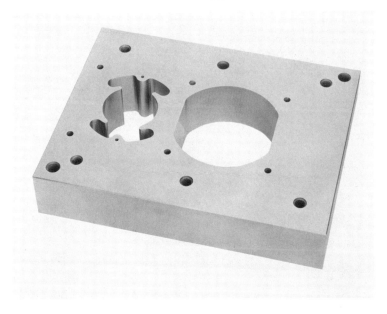

Fig. 7-10.   This complicated part, particularly the two large holes, demonstrate the versatility of the jig grinder. (*Moore Special Tool Co., Inc.*)

7-12) designs. The first uses either a mounted wheel, as shown, or a small straight wheel; the vertical grinder uses a type 6 or 11 cup wheel. Both can also be used with coated abrasive discs. Both, as is evident, are designed not for precision but for heavy stock removal.

The very small straight portable grinders, which are sometimes called die grinders, are used for all kinds of fine finishing, beveling, and touchup, particularly where total flexibility in positioning is important. The quality of the resulting job, of course, depends almost entirely on the skill of the operator, although much finer abrasives are of course used.

The portable grinder is, according to most industrial users, a useful utility tool. But in welding shops and foundries, on the one hand, and tool and die shops, on the other, it is part of the production sequence. In both instance there is usually adequate provision for storing and testing the tools. However, wherever portable grinders are used for utility work, there ought to be an ongoing effort by all management personnel involved to see that the tools are used safely (Fig. 7-13)—principally, that safety guards are in place—and that the tools are properly maintained, with adequate storage when they are not in use.

Portable grinders cause more accidents than their numbers justify, at least partially because everyone uses them and no one feels responsible for

Fig. 7-11.   Straight portable grinder with a cone mounted wheel, used for cleaning up a casting. (*Hitchcock Publishing Co.*)

their maintenance. But is *has* been difficult to design a wheel guard which is effective and light. Many operators feel that the guard gets in their way, so all too often, one of the first things they do with a new portable grinder is remove the wheel guard. If types 27 and 28 resinoid reinforced wheels can be used, it is a good idea to use them. Because of the reinforcement, they stand up under abuse better than vitrified or unreinforced resinoid wheels and thus have a greater factor of safety built in.

COATED ABRASIVE MACHINES

Belt-equipped coated abrasive machines make up by far the largest group of grinders using coated abrasives, primarily because they are the most productive of the group. The length of the belt provides a little time for the belt and the abrasive to cool off between passes across the piece-part; and even so brief an interval as this makes the belt more efficient. This may

Fig. 7-12. Vertical portable grinder with flaring cup wheel enables the operator to "bear down" on the work. This grinder operates at 12,500 sfpm. (*Hitchcock Publishing Co.*)

sound rather far-fetched, considering that a belt usually travels at a mile or more per minute, but there is significant cooling.

There are, however, a number of other types of coated abrasive machines.

Drum Sander

The drum sander, which is sometimes called either a cylinder sander or a drum grinder, consists of one or more cylinders or drums around which a strip of coated abrasive cloth is tightly wound. On one type, there is a slit in the drum parallel with its axis in which the two ends of the abrasive strip can be clamped. Another is designed so that a strip of coated abrasive cloth can be wound spirally around the drum. If marks on the finished surface can be tolerated, then the slit-drum type can be used, but if not, the spiral wound type is preferable.

Drums can be used on all kinds of materials, but mostly on nonmetallics. Those designed for cutdown and semifinishing can remove stock rapidly; the major problem is that there is so little time between passes for the abrasive to cool off. Drum sanders achieve tolerances on the order of 0.001 inch—not as good as can many belt types, and considerably less than can be ground with peripheral grinding wheels. They are well suited for large-volume flat-surface finishing of nonmetallics. If both sides of a piece-part

Fig. 7-13. It is good safety practice to run any grinding machine for a minute or so to check that everything is working properly before work is begun. This "bomb shelter," conveniently located on the work bench, can contain the wheel fragments in case of a breakage. (*Hitchcock Publishing Co.*)

are to be finished, it is possible to have a top and bottom finisher, or two machines in tandem, one to finish the top and the other, the bottom. Such a setup will handle stock at rates ranging from 12 to 72 linear feet per minute, depending on material and finish.

Coated Abrasive Disc Grinders

The first disc grinders were made by gluing a disc of coated abrasive material to either a vertical or horizontal disc. But these were single-disc grinders, the stationary models of which survive today mostly for wood sanding. Portable coated abrasive grinders, unlike their counterparts with abrasive wheels, are generally called disc grinders or sanders. The backing plates of stationary models tend to be hard and rigid; they are most often made of steel; but for portable tools the backing is most frequently resilient and probably slightly convex to ensure contact between the abrasive and concave contours. Resilience in the pad accomplishes two objectives: it distributes the pressure over a greater contact area than does a rigid backup and thus softens the abrasive action and aids feathering, which is essential in painting. The flexing of the pad also aids in getting rid of bits of work material which have become wedged between the abrasive grains, thus enabling the coated abrasive disc to cut more freely for a longer period of time.

Low-Pressure Coated Abrasive Machines

Coated abrasives find considerable use in what are more nearly sanding, finishing, and contouring applications rather than grinding. Four of these types are the pneumatic drum machines, slack-of-belt grinding setups, flap wheels, expanding-wheel units.

Pneumatic Drum Sanders. Pneumatic drums (Fig. 7–14) can be found on either stationary or portable grinders; the idea is the same in either type. A sleeve of appropriate size is slipped over a deflated drum, which is then inflated with air to the pressure desired; this can, of course, produce a fairly hard backup or, with low pressure, a soft abrasive surface that can cradle the piece-parts. A canvas cover over the drum will help protect it from wear, but the cover may possibly wrinkle at low air pressure. The disadvantage of this setup, as with all drum sanding, is the possibility of

Fig. 7–14. Air-operated drum used on a straight portable grinder with a coated abrasive sleeve. This is lighter than a bonded wheel of the same size; it adapts to slight irregularities of the surface; and the air pressure determines its hardness. (*Hitchcock Publishing Co.*)

heat buildup; but for the kind of polishing job for which pneumatic drums are best suited, it is less of a problem than is a situation in which stock removal is the prime consideration.

Slack-of-Belt Grinding. In any discussion of a machine tool using coated abrasives, some reference has usually been made to the backup roll, or platen, or perhaps to the magnetic chuck, which allows considerable pressure to be exerted on the abrasive and the piece-part. For the record, however, there is a type of coated abrasive belt grinding—or sanding or polishing—that requires no such backup—the slack-of-belt operations. The machine (Fig. 7–15) consists of two or three pulleys, one of which is powered. One part of belt stretched over the pulleys is vertical in front of the operator. In the operation, the operator simply holds in his hands the parts to be polished and presses them against the belt. The process is considered useful for final finishing of piece-parts where the only criterion is the appearance of the finished part but not dimensional or surface finish

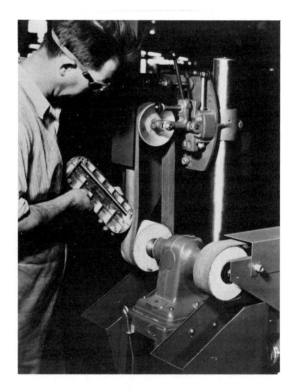

Fig. 7–15.   This unit illustrates slack-of-belt grinding or polishing. The belt adapts well to contours and rounded corners. (*Hitchcock Publishing Co.*)

tolerances. Chapter 11 considers the less labor-intensive ways of doing this work, ways which have reduced the need for slack-of-belt machines. However, for many applications, the fact that a worker can look at the part while it is still in process, and see when it is finished and looks good may well outweigh the labor saving from some other means of finishing.

Flap Wheels.   Another low-pressure coated abrasive use is the so-called flap wheel (Fig. 7–16) which consists of a hub built to hold sheets of coated abrasive cloth, often slit radially. When one of these is assembled and starts to turn, the strips of abrasive, held outward by centrifugal force, will polish and remove surface imperfections from anything held against them. A more aggressive action results when the strips of abrasive are backed up by brushes.

Flap wheels can do a better job of getting into corners or of finishing sharp contours than can slack-of-belt finishing, although there are obviously many applications where both can do equally well, so the choice of method would depend on which is available. Note that since a flap-wheel hub can be attached to virtually any electric motor that will turn a spindle, provided only that the weight of the hub is compatible with the capacity of the motor, a flap-wheel finisher is a less complicated machine to work with than is a slack-of-belt setup.

Fig. 7–16.   A mounted flap wheel of coated abrasive strips or sheets. This is wide enough to cover probably a 12-inch strip. (*Hitchcock Publishing Co.*)

Expanding Rubber Wheels.   An expanding rubber wheel could be described as actually an attachment for a portable hand tool or a small bench grinder rather than a separate type of grinder. The unit is a solid rubber wheel slotted from the periphery parallel to the wheel axis, and at a uniform angle from the radius. At rest, the circumference of the wheel is slightly less than that of the band to be used on it. After a band of coated abrasive has been slipped over the wheel and the machine is started, the wheel expands outward to lock the band in place.

SUMMARY

The first problem in writing about specialized grinding machines is no doubt to determine just when a type of machine becomes specialized; the second is to figure out where to stop, because the abrasive machining segment of the material-cutting industry probably has more variations of machines than has any other segment. One has to admire the ingenuity of the designers in coming up with variations of machines and even with variations on the variations. Quite a number of useful and effective machines available from a single manufacturer have not been included because of their limited availability.

In sum, where any plant or shop has a need for a machine or a combination of machines to perform some out-of-the-ordinary material-cutting job, there is almost certainly someone who has developed a machine to do just that job.

8
Bonded and Coated Abrasives—A Comparison

Only within the past 15 to 20 years could a comparison of bonded and coated abrasive machining even be considered. Up until at least the late 1950s, and to some extent even today, many users have felt that each has a particular use, with not very much overlap in applications. Currently there is not a definitive publication on the capabilities of coated abrasives, although there is a great deal written in the trade press. Only one or two of the few books that have been written about abrasives mention that there are abrasive belts and discs, and some otherwise thorough publications do not even acknowledge that they exist. But in one such comparison, put together by one of the trade magazines during the middle 1960s from comments by a panel of experts in each field, one of the proponents of coated abrasives had this to say, "Twenty years ago, had I been asked to answer questions such as yours, I would have had to say that we had made only small dents in a few areas. Today we are in practically every area of abrasive operations and in some of them have almost completely excluded bonded wheels."

It is essential to first define the limits of our comparison. Only the conventional abrasives, silicon carbide and aluminum oxide, will be considered. Neither diamond nor cubic boron nitride is used to any extent as a coated abrasive; and the only belts using these two abrasives are small and have only limited uses. One of the early things to be done is to specify the few areas of operation in which there is no competition between the two. Finally, since the principal competition is between bonded abrasive wheels and coated abrasive belts, it will be practical and convenient to refer to them simply as wheels and belts.

CHARACTERISTICS OF WHEELS

An abrasive wheel is an agglomeration of abrasive grain and a bonding material, both carefully chosen for a particular application, thoroughly mixed, fired at closely controlled temperatures, and then trimmed to finished size and sometimes to shape (shape is preferably established earlier in the manufacturing process). It thus has many layers of abrasive in random orientation, because the method of manufacture does not permit grain orientation to the best and most efficient angles, but enough grains are so oriented, and the abrasive wheel travels so fast, that the wheel is an efficient cutting tool.

Most of the abrasive grain in the wheel never gets a chance to cut, because only the outer third or so of the wheel is used before the wheel reaches stub size and is discarded. If the wheel is to maintain an efficient surface speed—similar to miles per hour—then its rate of rotation—revolutions per minute—must be increased as the wheel wears away.

Vitrified-bonded wheels, the bond type most often used in precision grinding, can be readily formed to almost any desired contour, limited basically by the type of form wheel dressing used and, on the small side, by the size of the grain. The maximum depth of any groove in the form is also affected by its width. It is difficult to "hold form" and maintain size if the cutting ridge in the wheel is both thin and high.

The width covered by any peripheral grinding wheel is determined by its thickness; for face grinding wheels, by diameter; and for segmental wheels, by the diameter of chuck holding the segments, which can provide a pretty wide swath. One-piece wheels can be molded to a maximum thickness of about 12 inches. Thicker wheels can be produced by cementing together two matched wheels, but then wheel weight can become a problem. Cylinder wheels in standard sizes peak out at about 24 inches diameter. And for segments, there is no limit caused by the molding process; each segment, of course, is well within the limits set above.

Theoretically, wheels can be produced in something over 100,000 different combinations of grain type and size, wheel hardness and structure, and bond. These are more combinations, in fact, than anyone has ever found uses for. And in actual practice, as was mentioned earlier, only a few will warrant serious consideration for any given application.

Finally, most abrasive wheels are noncompressible from the edge of the hole to the outside diameter (or from side to side for a face grinding wheel), which means that wheels are particularly well suited to produce close-tolerance work. Indeed, unless the wheel bearings are worn or there is some similar play somewhere in the wheelhead or the machine, there is no give worth mentioning. There has occasionally been discussion of manufacturing a more flexible wheel to compete more effectively with belts, but little has ever come of it.

COATED ABRASIVE BELTS

Coated abrasive belts—and indeed, all forms of coated abrasives—are still forced to live down the idea that sandpaper is good only for dry sanding of wood and other similar materials. Not until some time during the World War II era did waterproof paper (which permits the use of water and other liquids) enable coated abrasives to expand into machining metals, aided also by the development of heavier, more powerful machines, by improvements in the backings, and by refinements in manufacturing techniques, most prominently the switch to electrostatic coating.

A coated abrasive belt, then, consists of a backing material—most usually cloth, but sometimes plastic material or, for light-duty applications, paper—to which is adhered a layer of elongated abrasive grains held on end by two layers of adhesive, one applied before the abrasive coating and the other after. The abrasive coating may be full (called closed coat) or partial (open coat). Closed coating covers the entire surface; it is used for many common applications as well as for severe uses. Open coating covers 50 to 70 percent of the backing surface; it is used where chip buildup is likely to be a problem, because the spaces make it easy to get rid of anything that might become lodged between the abrasive grains.

Coated abrasives are manufactured in 4-foot widths and finished in lengths, usually quite long, to fit the available backing length. Length, of course, is not a problem. Nor is it a problem to make belts up to 4 feet in width; wider widths can be made by splicing, but the price rises steeply for widths wider than 4 feet. Inasmuch as such a width is sufficient to permit the sanding of plywood, or the descaling of practically any coil of sheet metal, demand for very wide sheets is not earthshaking. There are applications however, on which nothing else will do.

COMPARISONS

First, there are many applications where for one reason or another there is no competition between the two. For example, the grinding of wood, particularly in wide widths, is done virtually exclusively with abrasive belts.

Any application requiring the finishing of wide surfaces, particularly when the finishing is for the sake of appearance without regard to narrow tolerances, is also within the province of belts.

On the other hand, on very close precision work, to tolerances expressed in a few tenths (ten-thousandths of an inch) or perhaps in millionths of an inch, the rigidity of the bonded wheel generally gives it an edge over the other. One possible reason is the compression of the belt backing at the point where the abrasive comes in contact with the work. But there are still a host of applications with something like commercial or standard tolerances where one will serve as well as the other.

Grinding wheels have been termed the world's first throwaway tools, a designation that is probably accurate; but the abrasive belt deserves such a description even more so. Ideally the wheel is self-dressing; that is, as each grain becomes dull it is pulled out of the wheel surface and carried away by the coolant or the exhaust system. In practice, however, this doesn't happen too often, primarily because it is difficult to get that happy a match between the work material and the wheel. And many users prefer the extra wheel life that comes from a wheel that is slightly harder than the optimum, even though occasional dressing is required to renew the wheel's cutting surface.

A belt, on the other hand, is a true throwaway tool. It cuts best when it is new and its cutting performance declines in a fairly straight line from its initial level. There is no way to renew the cutting capability of a worn belt except by replacing it with a fresh one. However, a partially worn belt has better finishing capability than does a fresh one; it produces a better surface on the work; and on multihead machines, as has been suggested, belts can be moved from the first cutting head to intermediate heads and then to the final finishing head as they wear, so the performance of the whole machine is thereby enhanced.

Comparisons of wheels and belts for form grinding indicate that each has an area and the areas do not overlap too much. For example, if the form to be generated has sharp angles or corners, deep valleys, and small radii, it would probably have to be ground by a vitrified-bonded wheel. The length of the form is determined by machine size. It is always possible, in either surface or cylindrical grinding, to generate the form in steps, regardless of whether it is a series of teeth or a combination of angles and radii that repeats within the limits of the wheel's width. The pattern doesn't matter, as long as it can be reproduced on the periphery of the wheel.

Form grinding with belts, however, tends to run to shapes with larger radii and no sharp corners; these may well be roughed out by some other means and merely finished with the belt.

Costs

There is a kind of industrial maxim that claims that tool costs go up when coated abrasives are substituted for some other cutting tool. This may very well be true if only the cost of abrasive is considered, though that should not be the sole consideration. A wheel probably does not cut as well as a new sharp belt, because of the grain orientation on the belt. On the other hand, a conventional abrasive wheel has many more grains to cut with. For wheels, it is a matter of time needed for dressing and the loss of grain through dressing. For belts, it is a matter of time needed for belt changing and a probable straight-line decrease in cutting ability. But there have been some excellent reports of belts undercutting wheels in power cost per cubic inch of stock removed per minute. It is difficult to make a generalization; any cost comparison must be figured for each application.

Pressure

Pressure is essential for heavy stock removal, a factor that has rarely been a problem with wheels; but for years there was reason to question the ability of belt backings to stand up under the pressures of heavy stock removal. Within the past decade, however, the development of synthetic backings has pretty well put this idea to rest. This is not to say, of course, that a belt will never tear apart under pressure, but the strength of the backings—and, it might be added, the techniques of splicing and the adhesives—has improved significantly in a relatively short time.

Any comparison of wheels and belts must include some mention of the part that the contact wheel or roll plays in the performance of the belt. Whereas some plain rolls are used, mostly for fine polishing and occasionally for close-tolerance work, most contact rolls have spiral serrations, a combination of parallel alternating lands and grooves which greatly aids the machining action. In operation, the belt and the contact roll do not rotate in the same pattern, so that a given section of belt will strike the lands and grooves in some kind of alternating sequence. This variation in pressure helps keep the belt free of swarf, and probably lengthens its useful life. In fact, the design of contact rolls has become a very sophisticated science, some designs having as much as 80 percent grooves and only 20 percent solid lands on the roll or wheel face. This design increases the unit pressure per grain tremendously, provides good chip clearance, and pumps in additional air to keep the heat down despite the greater abrasive grain penetration. Different grains are in contact with each revolution, and their cutting plane is desirably reoriented. Fresh fracture facets are also exposed, to make the belt self-sharpening to a greater degree.

Materials

It is pretty well agreed that bonded wheels will cut harder materials than will belts, with the break point coming at about 52R$_c$. So belts find extensive use on all but the hardest carbon steels—on unhardened steels, gray and malleable cast iron, and nonferrous alloys. Some of these probably fall as well into the range for wheels, but the wheel range then extends to the hardest materials that silicon carbide and aluminum oxide can grind. The limit is set by the abrasive, not by the form in which it is used. This could pose a problem when some bimetallic parts are to be machined, but not the kind of problem that can be solved in a generalization. Actually there are some pretty hard materials that are ground with belts, and some quite soft ones that are ground with wheels, so there are other factors to be considered than simply the material.

COMBINATION OPERATIONS

Over the past few years there have been an increasing number of machines which have incorporated both wheels and belts to utilize the best qualities of both. This sometimes takes the form of teaming a wheel grinder ahead of a belt grinder; sometimes, as in belt centerless grinding, the belt removes the stock while the rubber bonded wheel regulates the rotational speed of the piece-part.

An example of this kind of operation is the snagging of billets in steel mills. This has been done for years with very coarse-grit resinoid wheels on swing-frame grinders, which grind across the billet. The procedure produces a coarse and often wavy surface on the billets, with the surface scratches running crosswise to the subsequent rolling operation. However, the introduction of a belt surface grinder operation after the wheel snagging refines the surface and changes the direction of the finish to be parallel with the rolling direction, thus producing a better-quality end product.

SUMMARY

This chapter is not intended to promote either abrasive wheels or belts as being a better machine tool. There are obviously a good many places where either one could be used as well as places where one or the other would be superior. And to reiterate a primary theme, there are many applications for which either one or the other would be superior to cutting tools, particularly when dimensional tolerances and finishes are more stringent than cutting tools can attain with ease, or when surface characteristics of the work make it difficult to use single-point tools. The material savings that result from better premachining operations that abrasives can take advantage of should not be underestimated either.

All this is not to say that wheels and belts are going to replace cutting tools completely, either now or in the near future. What is meant here is that there is a bigger place for both forms of abrasives than in simply cleaning up and finishing what some other machining method has started—which is what abrasive machining is all about.

9
Finishing with Abrasive Grain

The finishing processes using abrasive grain can be roughly divided into two groups: those in which parts are finished in quantity, known as mass or tub-type finishing; and those in which parts are finished one at a time.

A principal machine element in most mass-finishing operations is a tub or a container of some kind. These processes include vibratory finishing, barrel finishing or tumbling, spindle finishing, and considerable dry blasting. They involve substantial quantities of loose abrasive and some method of moving the piece-parts through the mass or stream of abrasive, usually called media. The first three finishing operations also include a fluid, usually a water-based mixture, called the compound.

It is convenient to refer to the other loose-grain finishing operations as single-part finishing, although there is no common term in use to describe the group. For this purpose, the group includes polishing and buffing. It might be argued that flat-lapping operations belong with this group also, because loose abrasive is involved. They are included in Chapter 5 with the other flat-surfacing operations because lapping is an operation concerned with precision rather than appearance only, as are polishing and buffing. Furthermore, lapping, or more particularly its related operation free-abrasive machining, involves some stock removal. And lapping machines have some resemblance to surface grinders, more so, at least, than the tub-type machines.

Blasting covers such a broad range that it is more practically discussed as a unit by itself. Most dry blasting involves quantities of parts, but very large parts can—often must—be finished one at a time. Automatic wet blasting is a high-production operation; but manual wet blasting is done one at a time.

214

MASS FINISHING

Barrel Finishing or Tumbling

Barrel finishing is probably the oldest of the abrasive processes, going back to mediaeval days when the knights' squires found out that they could tumble their masters' armor in a barrel with stones and sand and make the armor quite shiny and splendid. Today we have replaced the stones and sand with custom-made abrasives, carefully sized and sharp-cutting; we have replaced the simple keg or barrel, turned by a crank, with a neoprene-lined hexagonal or octagonal barrel turned by electricity and controllable by a timer, but we have not changed the principal action of the process. The cutting or finishing action still takes place as the parts and abrasive tumble together in what the trade terms the "cascade"—the fall of parts and abrasive when the spot where they are resting in the barrel reaches a level in its rotation at which they can no longer remain. Barrel finishing can remove external burrs left by previous processing. It can round off external corners and in a limited sense can remove stock (Fig. 9–1). There is also evidence that barrel finishing can remove some of the undesirable stresses in steel and replace them with compressive stresses favorable to improved fatigue strength.

Barrel finishing has several advantages. Although it is slow, it does not

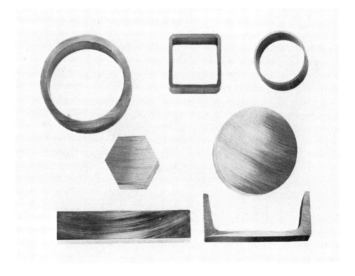

Fig. 9–1.   Small sampling of parts adaptable to barrel finishing. There are many others, of course. (*Hitchcock Publishing Co.*)

have to be monitored, and the time cycles are usually quite flexible. A timer can be installed, as noted above, or the cycle may run into an evening or night period when a janitor or a night watchman can turn off the machine. No harm is done if the machine is not unloaded at once. In short, aside from that needed for loading and unloading, the labor cost is minimal.

The process also does a superior job of finishing or cleaning up external surfaces of the piece-parts. Naturally, the finish produced and the time required are quite directly related to the size of the abrasive or media. The smaller the media, the better the finish and the longer the time cycle. Larger media does not produce as good a finish, but it does the job quicker. Barrel design is illustrated in Figs. 9-2 and 9-3.

However, the major benefit of barrel finishing may well be that it led to the development of vibratory finishing, to be discussed shortly, because barrel finishing *does* have some drawbacks.

For one thing, all the abrasive action takes place in the cascade, which is really only a small part of the total cycle time. For another, there is always the possibility that in falling the parts will nick each other, leaving flaws that the process will not remove. For a third, there is a possibility—quite probably more inconvenient than hazardous—of a gas buildup in the barrel, which of course must be tightly sealed. There is also the previously mentioned inability to do any work on the interior of the piece-parts. And

Fig. 9-2. Barrels are frequently designed like this, differing primarily in size. Each barrel is ordinarily a compartment, but it is possible to bolt two barrels together to make a two-compartment unit in one machine frame. (*Rampe Mfg. Co.*)

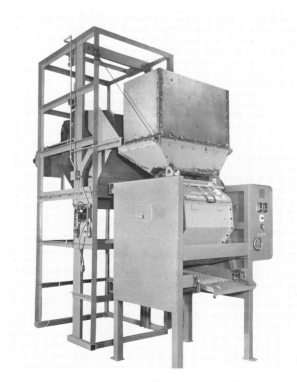

Fig. 9–3. A smaller barrel, but with auxiliary unit for loading. (*Rampe Mfg. Co.*)

since barrel finishing is a batch-type operation, it is not easily incorporated into a continuous sequence.

Separating the finished parts from the media after a run is a necessity, though rarely a problem. It may, however, require dumping the contents of the barrel into some kind of container with a screen on top, a screen which has openings small enough to retain the piece-parts and to let the abrasive fall through. Ferrous parts can be separated magnetically. The media is further screened so that the undersized grain can be discarded and the usable grain returned for the next load. Where the limitations of barrel finishing do not rule it out, the low labor cost can make it very attractive.

Vibratory Finishing

While barrel finishing is one of the oldest abrasive operations, dating from the Middle Ages, vibratory finishing is one of the newest, dating from only the middle of the twentieth century.

The idea of vibratory finishing is simple: if a tub containing abrasive or media and piece-parts is vibrated or shaken, the piece-parts will be worked on in a fashion similar to what happens in the cascading of parts and media in a barrel (Fig. 9–4). Burrs are removed, corners are rounded, scale and surface dirt are removed, and so on.

However, the vibrator represents a significant advance over the barrel. The action in the vibrator is continuous, not intermittent. Because the tub always is in an upright position, the top of the vibrator remains open during the entire processing cycle so that parts can be picked from the mass and examined at any time without stopping the machine or disturbing the other parts. There is less danger of part impingement and nicking in a vibrator, because the movement is engineered so that the parts proceed through the vibrator in a kind of pattern determined largely by the shape of the tub. This happens regardless of how the piece-parts are loaded into the tub. Of course, if part impingement cannot be tolerated, then some kind of fixture is needed to ensure that there is no contact.

The unique advantage of vibratory finishing is that in the process bits of abrasive work their way through any interior passages in the parts, deburring, cleaning, and radiusing the corners of interior surfaces and joints. There are only a few processes that have this capability, because interior

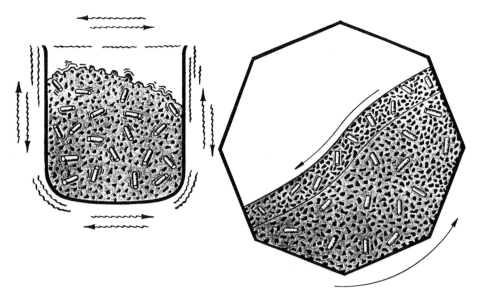

Fig. 9–4. Principles of vibratory (left) and barrel (right) finishing are shown in paired sketches of end views. Arrow in barrel indicates sliding layer of media and parts. (*Hitchcock Publishing Co.*)

surfaces are not accessible for most machining methods, and vibratory finishing is quite likely the best of those available, particularly for high production.

Design of Machines.   While there are of course many variations in the design of vibratory mechanisms, the principal visible difference in design is that some have rectangular tubs (Figs. 9–5 and 9–6) and others, doughnut-shaped round tubs (Fig. 9–7). Both types are widely used. With a rectangular tub, the maximum length of part that can be accommodated is something a little bit less than the length of the tub, although it could be pointed out that there are some open-end systems with rectangular vibrators where length ceases to be a factor. But the weight of the piece-part becomes a problem. In one major line of round vibrators, the maximum straight-part length ranges from 9 inches in the smallest (1-cubic-foot) model to 36 or 40 inches in the largest sizes.

Use in Finishing Systems.   Both designs are readily incorporated into finishing systems, with the necessary conveyors and other auxiliary equipment. The possibilities include setting up the system on an automatic batch-

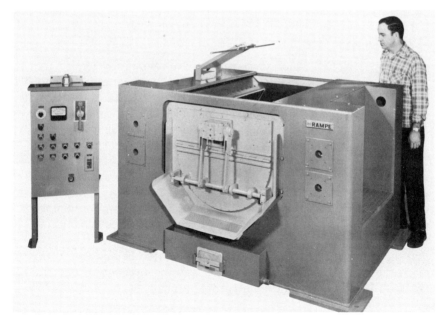

Fig. 9–5.   A rectangular vibratory finisher, with end discharge. Rectangular vibrators can be lengthened by bolting together two or more simpler units, each with its own mechanisms, to make room for longer work. (*Rampe Mfg. Co.*)

Fig. 9–6. This is a vibratory finishing system, with provision for recycling the usable media. It is a four-unit system. (*Almco Div., King-Seeley Thermos Co.*)

processing cycle, in which piece-parts are automatically loaded into the vibrator, processed for a set time, and then automatically unloaded to either a conveyor or a tote box of some kind for transfer to the next step. Another possibility, with straight-line processing, is to have the parts dropped into the entering side vibrator at predetermined intervals, processed en route through the vibrator, and discharged at the other end. At the discharge end, the parts pass over a coarse screen through which the media falls (Fig. 9-8). The media is then rescreened to eliminate the fines and recycled through the system. The fact that piece-parts can be examined during the vibrating process is of considerable help in determining cycle times and intensity of vibration. In the automatic circular vibrator (Fig. 9-9), as an example, one major manufacturer has designed a machine so that the vibrating cycle can be as vigorous as need be but also so that the discharge portion of the cycle is gentler and the direction of part travel is the opposite of the direction during processing.

Applications of Vibratory Finishing. Parts suitable for vibratory range from some not too much larger than the media to, with suitable fixturing, some parts of considerable weight and size (Figs. 9-10 and 9-11). It

Fig. 9-7.   A round vibratory finisher, designated by the builder as a "finishing mill." When the cycle is finished, parts and media cross the screen at the right. The parts are unloaded; the media remains in the unit. (*SWECO, Inc., Finishing Equipment Division.*)

has been estimated by one expert who originated much this type of equipment that fixtured parts are finished at up to four times the rate for unfixtured parts. The adaptability of the process to the finishing of interior surfaces has been mentioned. And, of course, the process is adaptable to any shape of part, as with barrel finishing.

Virtually all types of metals and metallic alloys are suitable for vibratory processing, as are many plastic and ceramic parts. Rubber piece-parts, for one example, can be deflashed if they are first frozen; at room temperature, the process is not so practical.

The vibrator is not (nor is the tumbling barrel) primarily a machine for stock removal, although some is removed in the finishing process.

It has been noted that abrasive action is continuous in a vibrator whereas in a barrel it is interrupted for any given part. In the barrel the abrasive action is confined to the cascade, that period when parts and media are falling. One interesting variation, which is somewhat of a compromise between the two, was a multiple-barrel unit, with the barrels mounted on a large faceplate that could be rotated, so the work and the media would be con-

Fig. 9–8. Separating parts and media on a vibrating screen. Media goes through; parts wind up in a tote box or in a conveyor. Further separation screens out undersized media. (*Hitchcock Publishing Co.*)

stantly sliding from end to end in the barrels. The developer claimed it could do everything that a vibrator could do but it required hours to do what the vibrator could do in minutes, and it could not be automated. And so it was eventually phased out.

Nonetheless, the mechanism of any barrel is much simpler than that of any vibrator. All that the barrel does is rotate at a relatively low speed. One does not have to be a mechanical engineer to understand that causing a load of perhaps several hundred pounds to vibrate at high rates is a strain on the whole driving mechanism. There may have been a time in the short history of the process when some of the drive mechanisms were not as

Fig. 9-9. An entirely automatic system for deburring rocker arms, controlled by the panel at the right. Parts circulate 14 minutes before being automatically discharged into the parts basket. Uniformity of finish reduces the need for deburring inspections, another saving. (*SWECO, Inc., Finishing Equipment Division.*)

reliable as they should have been, giving the process a negative reputation it did not deserve, but these faults have generally been corrected now, so the machines being produced today are both durable and efficient.

Spindle Finishing

Spindle finishing, sometimes called also spin finishing, is a loose-grain process which is sometimes nicknamed the "form-fitting grinding wheel." In this procedure, the piece-part(s), fixtured on one or more spindles, are lowered into a tub of rapidly rotating grain for a given period of time. The piece-part(s) revolve on the spindle to expose every surface uniformly to the abrasive action. Control of the process is achieved through variation of one or more of these major elements of the cycle:

Speed of the tub's rotation
Speed of rotation of the part on the spindle
Depth of submersion of the part
Angle of the spindle
Length of the cycle
Type and size of abrasive

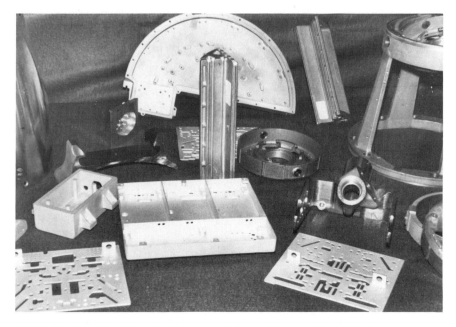

Fig. 9-10. Sampling of less than 5 percent of the total parts needing deburring suggests the range of deburring needs at one plant. Deburring is often a major problem. (*SWECO, Inc., Finishing Equipment Division.*)

These are all interrelated. There is a definite relationship between, for example, tub speed and work speed, though neither the two speeds nor their relationship are necessarily the same for two different types of piece-parts. Moreover, optimum speeds and relationships are things to be worked out by trial rather than data that can be read from a table.

The part must be covered by abrasive at all times. It is obvious that the abrasive will be thrown up in a kind of wave (like that in an ocean or lake) by the parts on the spindle and that behind the wave will be a depression. The depression behind the wave is termed the cavitation; the higher the tub speed, the deeper the cavitation—and consequently the deeper must the part be submerged in the tub (Fig. 9-12).

The angle of the spindle helps determine the degree to which the interior of the piece-parts is worked on. If the openings in a part are more nearly vertical, the abrasive action tends to be confined to the outside of the part. If the openings are horizontal, there is a considerable amount of internal abrasive action.

The effect of length of cycle is obvious.

The abrasive used in this operation is coarse fused aluminum oxide

Fig. 9-11.   Large cast aluminum parts weighing nearly 30 pounds each and about 1 cubic foot in size are deburred in a 20-cubic-foot capacity finishing mill. They are deburred in batches of six in a half hour cycle, compared with a hand deburring time of over one hour for each piece. (*SWECO, Inc., Finishing Equipment Division.*)

wetted down with water cut with periodic charges of detergent. The water flows in at the top of the tub and out near the bottom in a steady stream. The primary purpose is to keep the grain wet. The detergent softens the water for more effective flushing of sludge or swarf and of spent abrasive.

The process is practical for use with parts that are fragile, parts that need to be fixtured to prevent impingement, and parts for which close control of deburring and edge breaking is imperative. Part geometry is not a problem so long as the part can be chucked.

In spindle finishing, the tub usually rotates at speeds ranging from 275 to 600 sfpm, while the spindle normally turn at 8 to 36 rpm, rotating in a direction opposite to that of the tub. In this way, all the surfaces of the piece-parts are uniformly exposed to the abrasive flow.

BLASTING

Blasting is such a wide-ranging process in a number of ways that it is difficult to classify. It has been a means of cleaning large piece-parts (Fig. 9-13). At one end of the scale it can remove considerable amounts of scale and excess stock, although not, of course, to precision tolerances. With appropriate equipment it is a high-production operation, processing piece-

Fig. 9–12. A jet-engine-alloyed disc retracting from the rotating aluminum oxide mass after spindle finishing. The part spins counterclockwise to the rotating tub, at best point for maximum action. Cycles range from 15 seconds to 3 minutes, with no part-to-part contact. (*Almco Division, King-Seeley-Thermos Co.*)

parts in batches rather than on a continuous basis. At the fine end of the scale, it is a useful deburring operation, which can provide also a unique matte surface suitable for painting or coating. Blasting can also cut designs. And with a blast room or large blast cabinet plus an overhead trolley, blasting becomes almost an in-line process, with the operator merely loading and unloading parts. It can also be done one at a time, in a cabinet with arm openings and built-in gloves for the operator to hold the piece-part.

Blasting itself is the forcible "throwing" of an abrasive—steel shot or grit, aluminum oxide or silicon carbide, quartz, or even glass beads— against the surface(s) of piece-part(s). The propelling force can be air pressure, sometimes a vacuum, or, in what is called airless blasting, an impeller designed to rotate while projecting quantities of abrasive against the surfaces of the piece-parts. In the coarse range of abrasives the process is perforce a dry one; but if the abrasive is fine enough to be suspended in water, wet blasting is a possibility.

Fig. 9-13.   Traditional, and still extant, idea of abrasive blasting, which is needed for large units not easily moved. (*Hitchcock Publishing Co.*)

Dry Blasting

Dry blasting is usually done with steel abrasives—called shot in their original round state and grit after they break up—and impeller blasting, primarily for removing scale and rust. Coarse aluminum oxide and to a lesser degree silicon carbide are also used.

Steel shot weigh about twice as much per unit of volume as do either of the others and thus strike the work surface with greater force. They have no significant cutting action however; anything that is removed from the work surface is literally knocked off.

The weight differential has another effect. Because the manufactured abrasives are so much lighter, there are more particles per unit volume and they provide many more cutting edges. It goes without saying that the abrasives have to be tough and blocky to stand up to the pressure of impact.

There are at least two reasons for the continuing preference for steel shot and grit in dry blasting. One certainly is the original cost. Shot is definitely cheaper than manufactured abrasive. Another is that equipment has been developed largely with shot in mind as the abrasive, as have cycle times and other elements of the process. And it seems fairly obvious that there has not been sufficient discontent with shot to prompt active search for alternate cutting materials.

But there probably does exist evidence that would justify comparison on a larger scale. The greater cutting ability of manufactured abrasives should mean that cycle times can be shortened. It is quite probable that aluminum oxide breaks up faster than shot; but it may well be the case that the overall cost of aluminum oxide is equal to if not less than that of shot.

Manufactured abrasives are nonmetallic products unaffected by either atmosphere, alkalis, or acids. They will not cause rusting if they become imbedded in soft metals; in fact, impregnation may not take place.

In comparison with steel shot and grit for dry blasting, manufactured abrasives weigh less, cost more originally, and probably cut more effectively and thus shorten cycle time; they are not as durable, although they are probably more effective—but not as well accepted.

Wet Blasting

Since the abrasive used in wet blasting (Fig. 9–14) has to be of a size that will remain suspended in water with some agitation, it is obvious that this is a much different process than the dry blasting described just above. In fact, aside from the name and the use of essentially similar methods of propelling the abrasive, wet blasting and dry blasting are basically quite different processes. The following list of characteristics of the process will serve as a general guide to considering wet blasting as a finishing method:

1. Wet blasting is definitely an abrasive process, even though the use of 60-grit grain to powdery "flour" minimizes its stock-removal capability. Water also acts as a cushion on the abrasive cutting action.
2. Wet blasting will always produce a matte-type nonreflective finish; it is incapable of producing a mirror finish (Fig. 9–15). Some etching will always occur. The use of glass beads may improve the luster.
3. Wet blasting can be used on any hard surface—glass, ceramics, wood, metal; the harder the surface, the better. However, wet blasting is not effective on soft materials like rubber.
4. For finishing intersecting holes, slots or recesses, wet blasting has a general ricocheting or scouring effect, because the abrasive bits make more than one effective hit.
5. With the use of glass beads as media, wet blasting has proved capable of producing beneficial compressive stresses with resulting increases in

Fig. 9–14. Wet blasting system for cleaning wire. Wire unrolls from reel at right, is blasted clean as it passes through the cabinet, and is coiled up again on the left reel. (*Pressure Blast Mfg. Co., Inc.*)

fatigue life. Enough testing has been done to demonstrate that these desirable effects can be achieved without measurable stock removal.

The wet blasting process is always conducted in a completely enclosed cabinet (Fig. 9–16), which not only contains the spray of slurry but also permits the recycling of the water-abrasive mixture. In manual operation, now nearly obsolete, cabinets tend to be small and piece-parts are handled one at a time; but automated equipment may be more extensive, with conveyors carrying piece-parts past strategically located wet blasting nozzles. Some conveyors are built to rotate the piece-part as it rests on a flat table. Others have the piece-parts suspended on hooks, and possibly rotating also. Probably the limiting factor would be the weight of the parts; when they exceed either a size or a weight that is easy to move, it probably makes more sense to use a portable blasting gun. That, of course, would probably be a dry unit in some kind of enclosed room.

GRIT SIZE VERSUS SURFACE FINISH

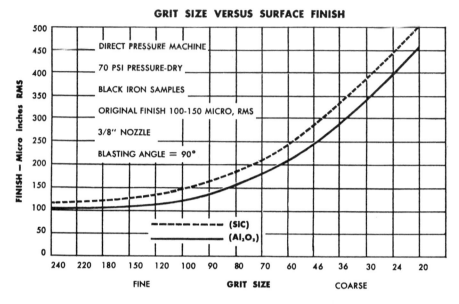

Fig. 9-15. Chart which relates grit size to finish indicates, as expected, that finer grain provides better finish than coarse grain, and that aluminum oxide consistently provides a better finish than does silicon carbide for any given size of grain. (*Hitchcock Publishing Company.*)

Fig. 9-16. A view from the inside of the cabinet of a setup for wet blast cleaning of printed circuit boards. (*Pressure Blast Mfg. Co., Inc.*)

SINGLE-PART FINISHING

There is no common term to describe a couple of finishing operations, used mostly to produce mirror finishes, that employ abrasive grain on the periphery of a relatively soft wheel. The two operations are called, sometimes interchangeably, buffing and polishing, with no clear difference between them. Strictly speaking, polishing uses abrasive grain imbedded in glue on the periphery of a leather or canvas wheel. It is the rougher of the two. In buffing, the grain is mixed with grease or wax and smeared on the wheel. Sometimes the carrier is thin enough so that the mixture can be sprayed onto the wheel.

Polishing

Not so many years ago it was a common sight to have workers in the grinding department "setting up" or "heading" polishing wheels. The operation called for the setup person to first roll the wheel in a narrow trough with hot glue in it, then roll the glued wheel in another trough containing abrasive grain of an appropriate size, and finally set the wheel on a rack to dry. After drying, the glue coat could be cracked in a pattern to provide more aggressive abrasive action.

If any considerable amount of production is involved, polishing with setup wheels would obviously require a considerable inventory of the wheels, so the result possible, taking the economy into account, is not always sufficient to counterbalance the inconvenience of renewing the abrasive layer—or of applying it in the first place. For these reasons, polishing wheels have been pretty much replaced by fine-grit polishing belts on soft contact rolls, slack-of-belt operations, by barrel or vibratory finishing, or sometimes even by bonded abrasive wheels. However, loose-grain polishing on setup wheels is as much an art as it is a science, so the ability of operators to achieve certain types of finishes may sometimes offset the more scientific approach of alternate methods.

Buffing

Buffing (Fig. 9-17) is a process using even finer abrasives—including materials like tripoli, rouge, or crocus—on even softer cloth wheels, for which reasons it frequently follows a polishing operation. Under some circumstances it may be the only finishing operation on such parts as automobile bumpers and other chrome-plated products. In fact, it is probably true that most of the buffing machines being produced today are custom-built machines designed for use in the manufacture of automobiles.

Fig. 9-17. Buffing a piece-part for appearance. Polishing with a cloth wheel is similar, with somewhat coarser grain. (*Hammond Machinery.*)

However, both buffing and polishing can be high-production operations, as shown in Figs. 9-18 and 9-19.

SUMMARY

Finishing in any machining sequence implies that the piece-parts have had practically all the excess stock removed, and that the only operation remaining is to bring the part to whatever surface finish and geometry is required. So in finishing with abrasive grain, in any of the methods mentioned above, there is no significant removal of stock, although it is, by definition, not stock removal work. A good deal of finishing is also for the improvement of the surface for the sake of its appearance; and deburring, if that is involved, is sometimes for appearance and sometimes for better operation of the part, particularly if the burr is on the inside where its coming loose could definitely be a hindrance to the part's proper operation.

The traditional method for finishing piece-parts is to do them one at a time (Fig. 9-17) by holding each against a buffing or polishing wheel or an abrasive belt—either the slack of the belt or the belt backed up by a soft contact wheel—with the belt on some kind of floorstand or bench-mounted

Fig. 9-18. Two-girl team loads and unloads pencil caps in a continuous rotary buffing and polishing machine. Timer meters buffing compound. Unloader wears cotton gloves to protect finished parts. (*Hitchcock Publishing Company.*)

polishing unit. But mass finishing, possibly in barrels, but more likely in vibrators, is changing that procedure.

One point of interest in mass finishing from the supplier side is worth noting. Most suppliers in the field prefer to provide their customers with a complete package—equipment, including conveyors and other auxiliary units; abrasive or media; and compound and/or liquids. This does not necessarily mean that they manufacture everything, but they prefer to supply all the elements needed, in which case they are unique. They are also very liberal in working out test samples so that they can recommend a complete package.

Most are basically equipment manufacturers. They buy abrasive for private labeling from one or more of the half-dozen or so major manufacturers of abrasive grain, and they buy fluids from a somewhat larger group of sources. This gives the customer one source to go back to, and the single supplier takes complete responsibility.

This approach is quite different from that in the grinding wheel and coated abrasive sections of the industry, where the machines tend to come

Fig. 9–19.   Inside view of the buffing and polishing operation. Pressurized spray gun for applying compound is in the center foreground. (*Hitchcock Publishing Co.*)

from one source, the wheels or belts from another, and the grinding fluids from a third. There are a few exceptions to this, but not many.

Loose-grain finishing, except vibratory finishing, is a relatively mature segment of the abrasives industry. Any process that requires one-at-a-time manual holding of piece-parts could certainly be questioned, except where the prestige or marketing value of hand finishing is substantial. Of course, at this stage of the machining process, the competition is pretty much between abrasive processes; other methods of machining generally do not have this level of capability.

It is entirely possible that vibratory finishing has the greatest growth potential of these finishing operations. It is by far the newest; it can be readily automated and incorporated into high-production lines; and it is not labor-intensive.

10
Grinding Fluids and Finishing Compounds

All machining operations generate heat, so practically all are done with some kind of liquid to combat the heat. Machining with abrasives is no exception. In fact, most abrasive machining operations where tolerances are involved are done wet, that is, with a coolant (the shop term) or a grinding fluid, the more inclusive and more recent term which recognizes that the fluid has other functions besides that of dissipating heat. The exceptions occur mostly in tool grinding and similar operations done by skilled machinists or tool grinders, where the need of the operator to observe what is going on in the tool contact area is regarded as being more important than the beneficial effects of the fluid. In tool rooms, there is also rarely the same push for productivity that there is in production grinding. It is very likely that all precision production grinding is done wet, whether the tool is a wheel or a belt. And it is worth noting that most loose-grain finishing is done with a fluid called the compound, although there is obviously less heat involved.

The first part of this chapter is devoted to grinding fluids or coolants used with abrasive wheels and belts; the end of the chapter discusses compounds used in the various loose-grain finishing operations.

GRINDING FLUIDS

Purposes

There appears to be little doubt that the optimum grinding fluid for a given job can increase production significantly when it is substituted for a less-

Fig. 10-1.   A common coolant setup for cylindrical grinding. It does a good job of washing away swarf but may not do so well in preventing loading of the wheel face. Its efficiency in controlling piece-part temperature depends on the volume that actually reaches the wheel-work interface. (*Bay State Abrasives, Dresser Industries.*)

suitable coolant, although one might be excused for wondering whether the improvements are as great as are sometimes reported. If nothing else about the operation is changed, and a change of fluid results in a significant jump in production, then that of course makes the story more believable.

However, grinding fluids are generally used for one or more of the following reasons, though not necessarily for all of them, nor in the order in which they are listed.

1. Fluids help reduce the temperature of the piece-part; they also help keep that temperature uniform, thus reducing the possibility of heat distortion and providing better control of size. (Fig. 10-1.)
2. Fluids tend to wash away the swarf—bits of metal or abrasive or any

other foreign matter—that might otherwise spoil the work surface (Figs. 10–2 and 10–3). And a good fluid ought to drop the bits of swarf readily as soon as it reaches the storage tank and not carry them around again to the tool contact area.

3. Fluids should reduce the possibility of loading the wheel or belt grinding surface with bits of metal. Such loading hinders the grinding action of the tool. Removing lodged bits of metal in some operations—crush form grinding is one—is so important that a special nozzle is often rigged behind the wheel to blast the swarf from its surface.

4. Fluids help reduce the friction between the abrasive tool and the piece-part, and to an extent, also reduce the resistance of the material to the abrasive action. This is basically a matter or lubrication.

5. Fluids help the wheel or the belt produce the required surface finish.

It follows then that almost any grinding fluid has to be something of a compromise, because the one that keeps the piece-part cool and clean (Fig. 10–4) is likely to be somewhat deficient in its lubricating potential; and conversely, the fluid that lubricates well is not likely to be effective in cooling

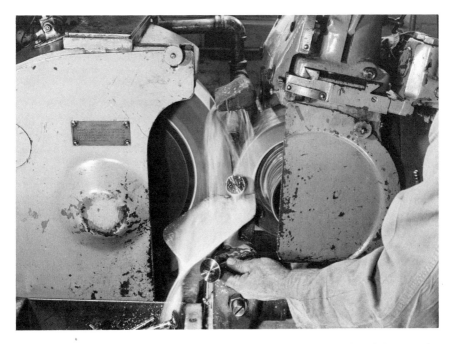

Fig. 10–2. On this centerless job there should be no problem with swarf, and there may be some coolant benefit from the two-wheel design. (*Hitchcock Publishing Co.*)

Fig. 10-3.   The problems and the solutions are much the same on the belt centerless grinder. (*Hitchcock Publishing Co.*)

and may retain some swarf in spite of the filters that are part of any good coolant system. Note also that any fluid must either be supplied in sufficient quantity so that it will cool off between trips through the grinding area or be cooled in some kind of refrigerating mechanism.

Central Systems

Some plants, for the most part those with a reasonably homogeneous mix of materials to be machined, find it practicable to install a central coolant system, using the same coolant for all or most of their machining operations. It is still possible, of course, to use some unit coolant systems where the central system coolant is distinctly unsuited for one or another of the machining operations. When it is possible to consider a central system, there are a number of economies possible in the purchasing, storage, and

Fig. 10-4. All the coolant spread out over the magnetic chuck of this surface grinder means very little if the nozzle under the wheel isn't carrying coolant to where the wheel meets the honeycomb. The breeze raised by the wheel should help carry coolant to where it should be. (*Hitchcock Publishing Co.*)

supply of coolant that can be realized. On the other hand, the possibility of casual contamination may be increased.

Unit Coolant Systems

The alternative to a central coolant system is, of course, a unit system, in which each grinder or a small group of similar grinders operates with a single coolant system, complete with its own filters, coolant chillers, and other elements, and functions independently of any other machines. Many grinding machines can be purchased with built-in coolant systems. There are many machines to which the system can be added on. The coolant handling system that comes as an integral part of a grinder is practically certain to have ample capacity to handle any production level of which the grinder is capable.

Unit systems have one advantage: the coolant can be selected to be the best match for the product and can be changed as the product changes. Such a system obviously offers a better chance for experimentation, something that is rather difficult in a central system holding perhaps several thousand gallons of fluid.

The disadvantage is that even a small buildup of swarf in the bottom of the coolant tank may be enough to reduce the volume to an inefficient level. As the volume is reduced, the coolant is circulated more frequently; it may not have time to cool sufficiently; and bits of abrasive and chips of the work material are more likely to be recirculated, causing random scratches, called fishtails, on the work surface.

Unit systems are most often used in small shops or in shops that handle a great divergence of materials. As a practical matter, it is likely that the care and cleaning of the system is more important for its efficient operation rather than the design of the system itself.

Properties of Grinding Fluids

The two principal functions of grinding fluids are cooling and lubrication, so needless to say, most grinding fluids are either water-based or oil-based. The fluid must provide good rust protection for ferrous metals, adequate wetting to keep the wheel face clean and free cutting, superior chip settling properties, resistance to rancidity, and, of course, safety for the operators. Moreover, the spent fluid must be of a composition that can be disposed of without environmental problems.

Water-based Fluids

Water has a number of characteristics that are useful in a coolant, but principally it is cheap and an effective cooling medium. On the other hand, water causes rust and has almost no lubricating capabilities at all. Therefore, water is nearly always modified by other liquids, such as the following, to make it more effective.

Soluble oils, which are light-base mineral oils, can be combined with water when emulsifiers and other chemicals are included to promote emulsification and stability. The general-purpose oils are used mostly on light grinding operations, whereas the high-performance oils, with considerable additions of extreme pressure additives, can be used on some pretty difficult grinding. These fluids cool better than straight oils, and don't lead to the smoking and misting difficulties of the straight oils.

The so-called semisynthetic fluids are usually mixtures of chemicals, emulsifiers, and no more than about a third (35 percent) of mineral oil. These are usually complex mixtures, the main advantages of which are that they are either transparent or translucent when added to water and they are effective in keeping grinding wheels and belts clean and free-cutting.

The synthetic fluids contain no oil at all, but they may be anything from water with rust preventives to some very complex blends of synthetic organic chemicals. Since there is no oil, emulsifiers are not needed, and performance is at worst rarely affected by the hardness of the water. These fluids are transparent, a feature that along with above-average rust control and cleanliness makes them well-suited for some operations which might well be done dry. It is probably fair to say that this is the area of the greatest rate of advance in grinding-fluid technology.

Straight Oils

It is generally easy to recognize a shop which has to use straight oils as grinding fluids; unless the exhaust system is very efficient, there is likely to be a film of oil on practically everything. This misting, plus a tendency of oil to make smoke on some severe operations, has tended to limit straight oils to jobs where nothing else will provide the lubrication that is needed. However, some progress has been made in producing nonmisting oils, and chillers can be used to accelerate the cooling of the fluid to minimize smoking. Oils are not outstanding, either, in their capacity to get rid of swarf, so the filtering system has to be somewhat more efficient—and the filters changed more often—than is the case with water-based coolants. With oils, however, less swarf is generally left in the settling tank.

Volume of Grinding Fluid

If the proceding discussion of grinding fluids seems less precise than the general subject of abrasive cutting, it is partly because there is still a lot to be learned by most users about why certain fluids are more effective than others. But, there are *two* aspects of the coolant problem which are, in comparison, crystal clear.

One concerns the concentration of the fluid, or the proportionate amounts of water and coolant concentrate in the mix. One of the most common ways to economize on coolant is to use less of the concentrate than the manufacturer recommends. This is usually false economy. This is to assert not that there are never applications where a thinner mix would be just as effective or perchance more effective than what is recommended, but only that the chances are rare.

The second has to do with the necessity of having the coolant that is pouring out of the nozzle in the vicinity of the abrasive-piece-part contact area really reaching that area. Just the fact that a flood of coolant is pouring out of the spout is no guarantee. The speed of the grinding wheel or belt often acts as a fan to blow the fluid away from the contact area, leaving the area virtually dry. Nozzles are available to negate this wind effect, but it's worthwhile to have some check on whether the nozzles on any grinders are actually wetting the contact area with fluid—something that cannot necessarily be determined by the quantity of fluid coming out of the nozzle.

With respect to the total amount of fluid available, the best suggestion is that any estimates should err on the side of too much rather than of not enough. And it's smart to make sure that any tank is cleaned out at regular intervals, so that any swarf that settles on the bottom of the tank is taken away; otherwise the swarf buildup may give a false impression of the volume of coolant available.

As for other cleaning methods, there are several. For any large-chip machining it is helpful to have a strainer ahead of the filter to take out the largest chips. Filters can take out particles down in the micron sizes, and they can be designed so that the filter fabric, which comes in a large roll, indexes to a clean new section whenever the section being used becomes sufficiently loaded. There are settling tanks, usually in central systems. There are magnetic separators (Fig. 10–5) to take out any tramp iron or steel. There are centrifuges which are particularly useful for the removal of

Fig. 10–5. A magnetic separator. Such a unit is frequently used where the recirculation of tramp iron or steel would be damaging to the finish of the piece-part. (*Hitchcock Publishing Co.*)

tramp oil (Fig. 10–6). And for heavy solids content in the fluid, the cyclone type of cleaner is recommended.

Actually, there are many applications where some combination of these types is desirable. Perhaps the most important thing to realize is that there must be enough clean coolant, applied so that it actually hits the contact area of the abrasives and the work, to make any high-production abrasive machining operation work up to its highest potential. And there are numerous ways of achieving this end.

Methods of Application

Whether the objective of the coolant is to cool the piece-part or to lubricate the grinding area, the point of the application mechanism is to deliver the fluid to the point where heat is generated—the point where the abrasive grains hit the work surface. The two most-used methods are the flood approach and the high-pressure jet at the back of the wheel away from the grinding area.

Fig. 10–6. Centrifugal separator. Dirty coolant comes from the machine through the hose in the lid and goes directly to the bottom of the bowl. Swarf is thrown against the bowl sides; clean coolant rises and goes out through the manifold at the right and then through the pipe to the tank. Capacity is a consistent 30 gpm. Some users buy two bowls to be able to clean a bowl without shutting down the machine for longer than it takes to change bowls. Very economical of space. (*Barrett Centrifugals.*)

Flood Application. Methods for flood application of coolants range all the way from open pipes above the grinding area to some sophisticated designs of nozzles, and there may be about as many of the former as of the latter. Most machines designed for wet grinding have a flattened nozzle that tends to spread the coolant across the grinding area (Fig. 10–7). There is of course also a system for collecting and recirculating the coolant.

However, with any type of nozzle there is still the possibility of "grinding dry with coolant" or, in other words, of having the coolant blown away by the fan effect of the rapidly rotating grinding wheel. This is particularly true in surface grinding. To remedy this situation, a number of special designs of coolant nozzles have been designed, one of which is shown in Fig. 10–8. Whether this is the final solution to the problem is of course a matter of opinion, but the principles it illustrates can be adapted in the design of a nozzle to fit many individual situations.

Blast Application. Blasting the wheel's grinding surface with a jet of coolant at some point away from the grinding contact area may not be considered technically as an application of a grinding fluid. The purpose, of course, is to keep the face of the grinding wheel clean, but hardly in the normal sense. The force of the jet dislodges the loose swarf, so it is practical to use the regular coolant that has been selected to aid the grinding ac-

Fig. 10–7. This is a good view of a flattened coolant nozzle which is as wide, maybe a little wider, than the wheel is thick. (*Hitchcock Publishing Company.*)

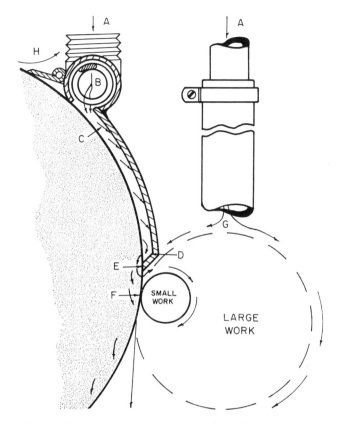

Fig. 10-8.   When heating is a real problem, a nozzle like this may be justified. It is rather more complex than most. It does not use high volumes of coolant. Fluid from pipe A flows into distributing manifold B. B and cover D must be adjusted to provide a full curtain of fluid at point E. This is for small work. For larger work, an extra pipe A distributing fluid from point G may be necessary, but at low volume and low pressure. The whole point is to get maximum cooling at grinding point F. (*Hitchcock Publishing Co.*)

tion and, in turn, prevent dilution of the coolant or the creation of a possibly undesirable mix.

Mist or Spray Application.   The application of coolant in a spray or mist is most practical on grinding machines basically designed for dry grinding, on which it is occasionally desirable to use a coolant. It is not common in production grinding. The spray nozzle—even a portable one— can be mounted on the grinder; only a small quantity of fluid is required; and there is no need for the splash guards, chip pans, and return hoses that are needed for flood coolant application. The fluid usually dries on the piece-part or can be easily wiped off.

However, using sprays or mists also requires a good exhaust system. Grinding coolants, especially the water-based kinds that would be used as mists, are not usually toxic, but there is no point in exposing operators even to nontoxic mists. Sometimes, though, an open fan may handle the ambient mist.

The purpose of a mist is for cooling only, and it is a surprisingly effective method. In the application, a small quantity of coolant is introduced into a high-velocity stream of air to form a mist. This stream is then directed into the cutting zone. Since it takes 100 times more heat to evaporate a given quantity of liquid than it does to raise its temperature 10°F, the small quantity of fluid that is evaporated does an effective job, particularly since the velocity of the air stream can carry it to the spot where it is most needed.

Through-the-Wheel Application. As has been noted, vitrified grinding wheels are porous, so it is possible to apply coolant through the grinding wheel (Fig. 10–9). This method requires a hollow spindle end and a gland

Fig. 10–9. Through-the-wheel coolant setup allows coolant to be fed through the hose into the special flange which directs it to the wheel arbor hole and thence out through the pores in the wheel. This is very effective in getting the coolant to the grinding point, but the pores in the wheel are easily clogged, and the inside of the wheel guard is sprayed along with the piece-part. Spray flies out of the wheel in all directions. (*DoALL Company*)

seal for transferring the fluid to the revolving spindle, from which the fluid passes by centrifugal force to the periphery of the wheel and thus to the grinding zone. This method ensures that the coolant will reach the contact area between the wheel and the work. Unfortunately, it also means that coolant is thrown off from the entire circumference of the wheel as well, thereby creating over the work area a mist that must be contained or exhausted. In addition to the special-design requirements, it is easy to plug up the small passages in the wheel with floating bits of swarf. For that reason, it is recommended that coolant applied in this manner be filtered through a 3-micron (0.0001-inch) filter. This method is not widely used.

MASS-FINISHING COMPOUNDS

The function of the compound in barrel and vibratory finishing has considerable similarity to that of the grinding fluid in abrasive wheel and belt grinding. Both have as a primary purpose the function of either keeping the abrasive grain cutting efficiently or of varying the grains' cutting properties. Both also serve to carry away the swarf that results from grinding.

In mass finishing, it should be noted, the term "compound" is sometimes used to designate the chemical formulation of the active ingredients and sometimes the mixture of these ingredients and water. The latter is the preferred definition because reference must otherwise be made to the "compound and water mixture, or solution," or whatever it is considered to be.

Contrary to the practice with grinding fluids, which are recirculated routinely, compounds are preferably used one time only. The trick in the selection is to find a compound whose active life is similar to the length of the run in the barrel or vibrator, and this is not always an easy task. But the alternatives are not attractive. If the compound is too strong, then there is waste as it is flushed down the drain. If it is too short-lived, then part of the cycle must be completed without benefit of compound, or else more compound is added during the cycle. In neither case are results predictable.

Types of Compounds

There are four general types of compounds—for cleaning, for cutting (with fine abrasive added), for burnishing, and for rustproofing.

Cleaning Compounds. Cleaning compounds, which are usually definitely alkaline or acid, usually have similar functions.

Alkaline cleaners. Alkaline cleaners must set up and maintain in the solution a safe pH to prevent rusting, pitting, and other forms of corrosion on the metal being finished. In addition, they should remove and keep in suspension any oils or dirt brought to the barrel or tub with the parts, remove and hold any swarf from the parts or media, and aid in removing these contaminants from the mass during the rinsing cycle. The alkalinity of the compound varies with different metals.

Acid cleaners. Acid cleaners perform essentially the same function as alkaline cleaners, but in addition are often used to remove heat-treat scale, rust, and other stains from the parts, as an alternate to pickling. Parts finished in this manner will generally have a better finish and color than abrasive-blasted or pickled parts.

After use of acid compounds on ferrous parts, however, the barrel must be quickly and thoroughly rinsed and neutralized immediately with an alkaline compound to prevent further action by the acid. Acid compounds can often be used to bleach or brighten parts which have become dark in color during previous processing. They can also remove any accidental abrasive impregnation. Acid compounds, however, must never be mixed with other cleaners.

Abrasive compounds. Abrasive compounds are essentially cleaning compounds fortified with additional abrasive grain for use with noncutting media such as steel slugs or with slow-cutting media to provide or speed up the cutting, or sometimes to create a better surface finish on flat surfaces of the parts, particularly when a finish in the low microinch range is needed. An abrasive compound is also added when an essentially self-tumbling operation is otherwise desirable.

Sometimes, however, it is more efficient simply to add a detergent compound to abrasive media to keep the media clean, open, and free-cutting.

Burnishing compounds. Burnishing compounds are probably the most complex of mass-finishing fluids. Since they are usually alkaline, they hold in suspension any dirt or oxides removed from the piece-parts. In addition, they help to restore the normal color of the parts and also provide some necessary lubrication.

Rust-proofing compounds. Rust-proofing compounds are used, of course, only for ferrous parts, and then only if the separation process is long enough to raise the possibility of rusting, or if the parts are either flat or saucer-shaped so that they cannot be effectively rust-proofed by a dip after separation. When separation is a long process, moreover, the burnishing compound which has picked up dirt may also discolor the parts if they are left in it too long. In such cases, it is necessary to rinse off the dirty compound before discoloration starts.

Rust proofing is most often done with a water-soluble oil, which in addition to coating the parts can also be used as a rinse. In instances where an oil might detract from the gloss or color of the parts, however, chemical compounds can be used.

SUMMARY

There can be no doubt that the use of a coolant or grinding fluid is beneficial in many, if not most, abrasive machining operations. This has been recognized for well over 100 years, ever since farmers discovered that they could do a better job of sharpening scythes and other farm tools on their grindstones if they hung a bucket of water with a small hole in the bottom over the wheel so that the water would dribble down and keep the wheel wet.

We've come a long way from that, but not so far that many individuals are able to design coolants from scratch to perform efficiently in particular situations. From experience gained in perhaps thousands of applications, suppliers and some knowledgeable users can do a creditable job of selecting coolants, but their knowledge does not necessarily apply in new situations nor on new materials. Selection is still too often a matter of trial and error.

Of course, if any coolant comparison tests are contemplated, it is vital that everything except the two coolants being compared be kept as identical as possible.

Compounds in vibratory or barrel finishing are also necessary, but the selection of the compound(s), the number to be used, the rinsing procedures, plus the time that each is to be used, are all matters usually best handled by someone with experience. It is always possible to get a second opinion.

11
Abrasive Machining versus Cutting-Tool Machining

Any comparison of abrasive ("small-chip") machining with cutting-tool ("big-chip") machining has to start with some basic ideas about the processes and the tools. First, there are some areas where the two do not compete at all, though there is no consensus about the boundaries. And the choice of one over the other may rest, for adequate reasons, on factors other than demonstrable differences. For example, steels harder than R_c 55—give or take a little—are generally ground. But a shop with a powerful and modern milling machine would probably push a few points to the hard side, while another with a 250-hp surface grinder might grind some considerably softer steels on it. But hard materials are generally ground.

Where there is considerable stock to be removed from the piece-parts, the removal is generally done with a cutting tool. But there is a difference of opinion about what constitutes "considerable." Not too long ago it was thought to be 1/16 inch; then it became 1/8 inch, then 1/4 inch, and in many shops there would be no hesitancy about grinding off 1/2 inch, certainly in surface grinding on a vertical spindle machine. But somewhere in that range someone might—or should—begin questioning how much stock is really needed, and whether the part might be better ground because there was less stock to take off. This would be the case usually if the extra "overcoat" of stock could be eliminated earlier.

FACTORS IN TOOL PERFORMANCE

There are a number of factors to be considered in comparing tool performance, which is the heart of any machining operation. These include hard-

250

ness of the tools, heat sensitivity, depth of cut or stock-removal capability, which ought to be considered along with the amount of stock which the tool needs to function efficiently, design of the part, the amount of fixturing that is needed to hold the work, and the tool's capability of doing the whole job, what might also be called its range of use.

Tool Hardness

As a general rule, any cutter should be harder than the material it cuts. For while the tool is cutting the work, the work is, in a sense, also cutting the tool, which is the cause of tool wear. Also, as a general rule, the greater the difference in hardness, the better the tool will cut.

For metal cutting tools, the hardness difference is usually not great, a condition which increases tool wear and shortens the intervals between resharpenings, a job that is usually done off the production floor, thus requiring transportation time and expense. Throwaway carbide tools, however, have eliminated this to some degree.

For abrasive wheels the bits of abrasive are always significantly harder than the work material—diamonds in relation to carbides, cubic boron nitride in relation to alloy and tool steels; aluminum oxide relative to most other steels, and silicon carbide relative to most nonferrous metals and nonmetals, the customary materials on which the abrasive tools are used.

It is worth noting, parenthetically, that hardness alone does not make a material a good abrasive. Several quite hard materials have been tried over the years in the search for better abrasives, but all except diamond and cubic boron nitride have proved not to be successful for general use.

Resharpening (dressing) of abrasive wheels is probably less frequent than resharpening of metal cutting tools, but more to the point, it does not require the removal of the wheel from the machine. The wheel can be dressed frequently without even interrupting production.

Abrasive belts are used until all the grain is dull and then discarded.

Loose abrasive grain that has become dull is usually undersized; it can be screened out and discarded prior to recycling.

Resistance to Heat

Heat is, of course, the arch-enemy of all machining. The degree of heat that the tool can tolerate without losing efficiency is a major factor in establishing work cycles.

As was pointed out in Chapter 10, one of the principal reasons for using cutting or grinding fluids is to keep the work and the tool from getting too hot.

Generally, neither the abrasive grains nor the bonds or adhesives that hold them are particularly heat-sensitive. After all, the major abrasives are crystallized at temperatures well above the melting point of any steel. The temperatures at which vitrified wheels are fired run between 2000 and 2500°F; and resinoid wheels, even though they may be formed at the relatively low temperatures needed to solidify the thermosetting bonds, can withstand the high temperatures of foundry and billet snagging without problems. One wheel-making company even explored the possibility of cutting steel rod with an abrasive wheel while the rod was still red hot, and the attempt was reported successful with respect to the wheel. The need for temperature control in any abrasive machining operation is for the benefit of the piece-parts, not the wheel or belt.

Depth of Cut and Stock Removal

Abrasives of any kind cannot match the depth of cut per pass of cutting tools, but the amount of stock removed per unit of time is quite another matter. The chips produced by abrasives are microscopic in comparison, but there are simply so many of them. In many situations, probably most clearly in vertical-spindle surface grinding with wheels, and in wide-belt grinding, even though the latter is a line contact, abrasives can match, and often exceed, the total volume of chips produced by, say, a milling cutter under similar circumstances. Upon the introduction of a high-powered new belt surface grinder a few years back, the machine was reported to remove, with its two 30-inch-wide belts, 600 pounds of cast iron per minute. A major metalworking manufacturer bought two of these machines in preference to the five milling machines their engineers had determined would be needed for the job.

There is a paradox in the business of stock removal. While it is true that a cutting tool is more efficient in removing stock in terms of energy used, (horsepower per cubic inch per minute), it is also true that the cutting tool, to be efficient, has to have more stock to remove. This is particularly true if there is any scale on the surface of the piece-part, or if there is any surface hardening. The surface must also be continuous, without any slots or breaks of any kind, for best cutting tool operation. An abrasive wheel or belt, however, cuts with equal ease through hard or soft spots and bridges interruptions in the surface without any difficulty. And either wheel or belt will cut as thin a layer as is needed; this kind of operation has been the domain of abrasives for years. The end result is that any extra cost in energy is often recovered in savings in material. After all, there is no point in adding material to a piece-part, at any point in its processing, just so that the excess can be taken off later. Abrasive machining may use more power, which is more important now than it once was, but the total cost of the end product is often less because of savings in material, to say nothing of possible savings in labor, transportation, space, and other cost items.

Design of the Part

This brings us to another consideration: the original design of the part and two salient features of the design. The first one is the amount of stock that is left for machining, and the second, the amount of the area to be machined. In other words, if a part has been designed for cutter-tool machining, it may not do much good to switch the part over to abrasive machining without adjusting the design of the part.

If the designer routinely assigns to the part the allowances that are necessary for cutting tools, there is little to be gained from machining it with abrasives. However, with abrasives, the controlling element is not the amount of stock needed to make machining efficient but the capability of the preceding operations in coming close to the dimensions of the finished part. That capability is being improved constantly as foundry or forging techniques improve.

The size of the area to be machined is another critical factor in part design. A cutting tool works best on a continuous surface, and if the part were, say, a motor base such as is sketched at the top in Fig. 11-2, it would be better to design it with a flat top and drill the bolt holes after milling. However, what is actually needed is the frame sketched at the left in the same figure, with the opening in the center and slots for the bolts all finished before grinding. The saving in material is evident, and the part would be finished after it left the surface grinder (Figs. 11-1 and 11-2).

So there can be developed three prime requirements for designs that favor grinding:

1. Surfaces to be finished must be made accessible to the grinding wheel or belt.
2. The amount of metal to be removed should be held to the lowest possible figure.
3. The areas of plane surfaces to be finished should be kept as small as possible.

The first two of these would apply to any kind of part; the third is obviously of concern in surface or disc grinding.

Fixturing or Workholding

The fixturing requirements of different machining processes is a factor in their comparison, partly because of the cost of the fixtures, but primarily

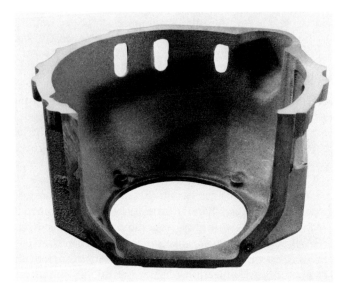

Fig. 11-1. Grinding the top rim of this torque converter housing is an excellent example of a part well suited for abrasive machining on either a vertical-spindle, rotary-table surface grinder or quite possibly on a big belt surface grinder. It would be extremely difficult to mill. Principal points favoring the choice would be the small area to be machined and the large void in the surface. (*Hitchcock Publishing Co.*)

because of the time involved. The element that makes the difference is the amount of pressure that the tool exerts on the work.

For instance, any flat-surface machining on a milling machine or a planer or shaper involves considerable pressure, usually enough so that the piece-part has to be clamped, thus increasing setup time and cost. However, pressure with a grinding wheel or belt on a surface grinder is rarely enough to require clamping. Most work on surface grinders can be held on flat magnetic chucks; indeed, this is probably the only machine tool of consequence that can get by with only the holding power of a magnetic chuck. All that is usually needed for most ferrous work is to set the piece-part(s) carefully on the clean chuck; and most nonferrous work can be held between steel retaining blocks. If the surface to be machined is not parallel to the chuck, an angle block can be used to make the surface parallel, as it has to be. Clamps, if needed, are simple and fast.

For belt surface grinders, the magnetic chuck is also used, along with angle blocks. A considerable amount of belt grinding is done using conveyor belts, sometimes with and sometimes without form-fitting pockets.

The point is that there is little if any problem in fixturing flat-surface work for either wheel or belt grinding. And in disc grinding, fixturing often

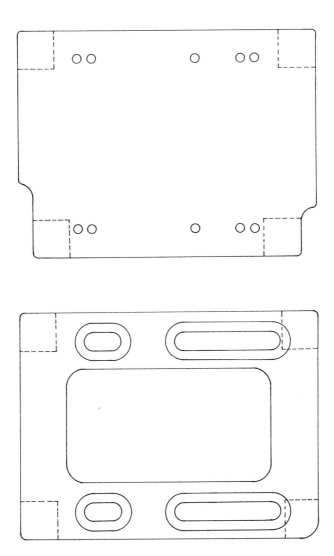

Fig. 11-2.   This dual sketch shows (top) the way the machine base was originally designed for milling and (below) the way it was designed for abrasive machining. For milling, the whole top area is machined before drilling the 10 holes. For grinding, only the pads around the slots need to be machined, after the other work on the base is complete. The reduction in machining area is reported to be about 95 percent, and there is a significant reduction in material with the cutout area in the center of the part. (*Grinding Wheel Institute.*)

comes down to a load-unload sequence, with only loading as a constant requirement, though parts may be unloaded by gravity.

Between-centers machining of shafts and similar shapes, with either abrasives or cutting tools, requires similar and in many cases identical workholding. There may be a slight edge for the abrasive tool, but for practical purposes it is a stand-off. In centerless grinding there is, however, no part setup time for individual pieces; the necessary adjustments are made at the beginning of a run of parts and they usually hold throughout the run.

Tool Range

Tool range is a comparison factor which can best be described as the portion of the machining sequence on a piece-part which can be efficiently done by a given tool. The roughing portion of the sequence, which involves stock removal, has traditionally been done by cutting tools, and the finishing to precision tolerances, by grinding wheels or belts, or sometimes, by loose grain. Each of the tools in question—milling cutters and lathe tools on the one hand, bonded abrasive wheels and shapes, or coated abrasive belts on the other—has some areas which are traditionally its own; each has areas in which it does not compete at all. For example, few will question that for very close dimensional tolerances and for low microinch surface finishes, some form of abrasive machining is needed; the requirements are simply beyond the capability of any cutting tool.

But differences of opinion begin to arise when the question of stock removal with abrasives is brought up. Is the limit on the order of 1/16 to 1/8 inch or does it range up toward 1/2 inch? It is granted that the hardness of the work material must be considered. Removing 1/2 inch of cast iron may be easier than removing 1/16 inch of, say, many alloys. But then one can raise the question, as has been done earlier, of the necessity of the extra stock. It is difficult to justify adding material to a piece-part just so that later it, with a little more, can be removed.

Producing Formed Parts

The ability of a peripheral grinding wheel to produce formed parts is difficult to overemphasize. In case after case, some of which will be noted later, formed grinding wheels have been tried on such piece-parts only after milling had been tried and found wanting, either from the standpoint of part quality or production level.

One of the simplest of this type of application is a lengthwise slot, which needs only a wheel of appropriate thickness for the width of the slot. Or if there are several slots parallel to each other, several wheels can be ganged on the spindle with spacers of appropriate size between. Or, of course, the grinding face of the wheel can be crush-dressed or diamond-dressed to the reverse of the form in the work. And simple concave or convex radius faces are readily dressed with a diamond radius dresser.

Total Processing Cost

Brief mention must be made of the folly of depending on tool cost as the sole deciding factor between abrasive and cutting-tool machining. Even when the total cost of abrasives is greater than that of cutting tools, the total cost of the finished part is frequently less with abrasives.

Milling versus Vertical-Spindle Surface Grinding

This discussion can be set forth plainly in the following excerpt of an article summarizing a comparison of milling and surface grinding. It applies pretty much across the board for cutting tools and grinding wheels.

Milling (1)	Grinding (2)
Avoid sudden changes in cross section. Although this is considered no more than ordinary good design practice, it is mentioned here because of its effect on the machinability of the casting. Thin sections, which cool more quickly, can result in chill spots which are of higher hardness than adjacent sections. Such abrupt variations in hardness can have disastrous effects on the cemented carbide cutting edge.	**Chill spots in castings are not of concern** on a vertical-spindle surface grinder. They can have a considerable variation in hardness without causing trouble. Uniform sections and uniform hardness are not abrasive machining requirements. The grinding wheel doesn't care whether the part is hard or soft. In fact, hard and soft spots both help the grinding wheel to dress itself.
Provide enough material for a good machining cut. When the tool must make a scale cut, the effective life of the cutting edge is reduced. You do not always reduce machining time when you leave little stock; you may in fact be losing time and money through excessive cutting edge wear. Try to leave enough material so that the nose of the tool is always under the scale.	**Why pay to put on stock and then turn around and pay to take it off again?** Abrasive machining requires only enough stock to assure that the part can be ground to the desired geometry. Sometimes it may not even be necessary to remove all the scale as long as the surface is solid. This saves material in addition to the savings from not removing stock that is not needed on the casting.

Source:
(1) Reprinted with permission of the Society of Manufacturing Engineers, Dearborn, Michigan, from "How to Avoid Problems in Machining Iron Castings," by Alfred M. Thomson.
(2) Reprinted by permission of Hitchcock Publishing Company.

Milling	Grinding

Avoid interrupted cuts. An interrupted cut is one where the tool alternately enters and leaves the workpiece during the cut. Because cast iron has relatively low tensile strength, the tendency of the cutting forces is to break out the edge of the casting as the tool leaves the work. Although the interrupted cut is hard on the tool, a suitable tool material capable of withstanding impact shock loading can be selected. A side effect is the potential production of thermal cracks in the tool material.

Interrupted cuts are exactly what the grinding wheel needs to help in self-dressing. There is no need to baby the wheel, and no concern about edges in the casting. This type of surface allows for easy application of the coolant, either through the wheel or from the outside.

Avoid thin-walled sections. Thin-walled castings are difficult to deal with. They are often hard to hold in chucks and fixtures. During machining, the thin walls tend to produce high frequency vibration. The result of this vibration can be seen as "chatter marks" on the workpiece and as chipped or broken tools.

Of course everybody should try to avoid thin-walled sections, but there are times when design characteristics won't allow it. The rigidity of the grinder, using the minimum feed available, can help adapt grinding technique to the part. It is unusual to have to consider vibration of the thin sections. If vibrations do occur in roughing it is usually possible to remove any resulting traces of chatter in the sparkout passes.

Remember a casting has draft. This makes the depth of cut vary when cutting with the draft. Make sure that the machine has enough horsepower and the workholders sufficient strength to handle the greatest depth of cut which will be encountered. This is particularly important when machining to a corner or shoulder. Stalling of the machine could result in considerable damage to the workpiece, the cutting tool, and the machine tool itself.

Draft is not a problem in abrasive machining. With the grinder there is a continuous downfeed because the table is constantly rotating or reciprocating under a continuous downfeed of the grinding wheel. Thus you always hit the high point of the work first, anyway, and feed gradually into the full surface area.

Consider proper application of coolant. Although many cast iron machining applications can be performed satisfactorily in the dry condition, properly-

Too bad that the cemented carbide tool is so sensitive to coolant. Coolants are a must for abrasive machining for many reasons, one of the principal ones being

Milling	Grinding
applied cutting fluids can extend tool life. Cutting fluids are best applied under pressure directly at the tool-chip interface. The use of coolant is particularly dangerous, however, in interrupted cut applications because it may promote the formation of heat checks in the edge of the cemented carbide tool.	to keep the work cool. Coolants on high-powered vertical-spindle surface grinders provide better cutting action and longer wheel life. This is documented. Finally, the coolant absolutely carries away all the chips and swarf.
Have the tool enter and leave the work gradually. This is a machining technique that can aid in avoiding heat checks and chipping out of the casting. The use of rounded corners and gradual section changes in the design of the casting can help in promoting this technique. Tool geometry can help in achieving this objective. Use of a high lead angle helps to avoid abrupt entry and exit.	**There is total wheel contact** in abrasive machining on a vertical spindle grinder, and line contact on any type of grinder. There is nothing like cutter vibration or tooth-load. The direction of wheel pressure is down on the part, toward the magnetic chuck, which tends to hold the part tight in relation to the cutting action of the wheel. Entry and exit angles are not critical.
Keep the castings in compression while machining. Although iron castings are relatively low in tensile strength, they possess relatively high compressive strength. Take advantage of this characteristic wherever possible by machining from the outside in, so that tool forces are directed toward solid metal in the casting. Finishing the cut in solid metal also avoids the tendency to pull metal from the casting.	**In abrasive machining, no compromise** needs to be made to favor the design of the piece-part by cutting in any particular direction. In comparison with milling, the horizontal forces of grinding are much less, though the downward pressure may be as great or greater. So, while it makes sense to support any projections from a piece-part, where the design of the part does not provide adequate strength, there is no need to be concerned about the direction of the cut.
Select compatible clamping and holding methods. This is particularly important when machining malleable and modular iron castings, because they may deflect when force is applied. When a casting is machined in a deflected condition, it could be dimensionally out of tolerance	**The magnetic chuck on a surface grinder is compatible** with practically any design of iron casting, and the extra fixturing to accommodate odd geometry is usually simple. Castings can be roughed with maximum flux on the chucks, and after roughing the flux can

Milling	Grinding
when it is unclamped. Be sure you support the casting well. This may require a special fixture, such as a pot-type retainer or one with spring-loaded snapover clamps.	be reduced at sparkout, thus relieving any deflection that would come from a distorted part. Force sufficient to distort a part is unusual. Often some simple blocking is sufficient for support.

Guard against high-frequency vibration. As has already been mentioned, a potential problem in machining thin-walled castings is chatter. Chatter at the tool cutting edge usually results in irregular machining of the workpiece and possible tool failure.

Where a thin wall section is necessary, the tool engineer may compensate by filling the casting with some liquid and allowing it to solidify before machining. Another technique is to use a spring-loaded damping device. A third alternative is the use of positive rake tools with light cuts.

In abrasive machining vibration is a problem, as it is in any other machining operation. We also try to contain vibration, which helps to deteriorate the cutting action of the abrasive and increases the rate of wear. Chatter is less likely to occur on a rigid machine with light downfeed. Filling the casting with a material that solidifies at operating temperatures and melts at a slightly higher temperature is often practical, when the section cannot be ground otherwise.

Use heavy feeds on scale cuts. This is the machining technique designed to offset the need for running the cutting edge in scale. A heavy feed forces the chip up and out ahead of the cutting edge, breaking away the scale before the cutting edge reaches it. In this way, tool life is substantially extended. This is one of the reasons it is essential that sufficient machining stock be provided in the design of the casting.

Behind the smoke screen, this is another suggestion for designing in more stock so that it may be taken off. Abrasive machining requires only enough stock to ensure the surface required to do the job. In many instances, as long as the surface has the required geometry, it may still have some shine spots where the scale was not entirely removed and still be entirely satisfactory.

In summary, compared with the single-point cutting tool, the grinding wheel . . .
1. Is less sensitive to hardness variations in the work material.
2. Is more economical in materials, since it requires less stock for machining.
3. Is not affected by interrupted cuts, and in fact, performs better for them.
4. Adjusts better to thin-walled sections.
5. Is not affected by either draft in castings or tool entry or exit angles.
6. Requires less complicated workholding devices.

Why Not the Midrange?

One further observation on the stock-removal capabilities of the grinding wheel is in order, even though it may be only tenuously related to the comparison of wheels and cutting tools.

There is virtually no opposition to the use of grinding wheels where either precision tolerances or surface finishes of excellent quality are required. In fact, the grinding wheels of a century ago were used for just this purpose.

Moreover, abrasive grinding wheels are also preferred in foundries and steel mills for snagging of castings and billets, where the only criterion is the ability to remove excess metal as rapidly and as economically as possible. In fact, it is doubtful if cutting tools were ever seriously considered for such applications; at least they have not been so considered for a long time.

It is true that precision grinding is done with vitrified-bonded wheels, and snagging, with coarse-grit resinoid-bonded wheels. And it is also true that neither precision machining shops nor foundries and steel mills have, until the past few years, been extensively informed about what was developing in the other area. But then, for a long time vitrified wheels were used for snagging and not generally phased out of that application until after World War II. Finally, foundries do no precision grinding at all and machine shops rarely get into very heavy stock removal. But the question remains: why not abrasives for primary machining—often called roughing—operations?

CASE STUDIES IN STOCK REMOVAL AND FINISHING

Case studies as a means of demonstrating the superiority of one method of machining come somewhat under a cloud because they are rarely fully documented and because it is not always possible to ascertain whether the displaced method they report is the best that it could possibly be. But a production floor is not a research department, and production decisions cannot always wait until all the facts are in. Furthermore, not all shops and departments go into a detailed breakdown of cost and production factors before making a change, and those that do may be reluctant to publicize the figures. And, of course, there are situations where factors other than the raw comparison of the application enter into the decision. Then there are those instances where the equipment available simply is not doing the job, so some other machinery or some other method must be used.

But any measure is better than no measure at all; although the ones which follow are not as detailed as would be desirable, they are all authentic, and, so far as can be determined, represent definite savings.

BONDED ABRASIVE WHEELS

The following reports detail the results obtained by shifting an operation done with some kind of cutting tool to one done with a bonded abrasive wheel. Wherever possible, the data have been converted to percentages of change, which, for most readers, is more meaningful. Wage rates and machine costs vary considerably from plant to plant, but the magnitude of most of these changes is such that they represent substantial savings even when discounted.

Aircraft Roller-Bearing Blanks

The blanks were being cut, two at a time, on screw machines. When the cutting was shifted to an automatic abrasive cutoff machine, it could handle eight cuts at a time. This was, of course, the big saving. The saving in the cutting operation came to nearly 70 percent of the previous cost, and the total cost of the finished rollers dropped 45 percent.

The blanks were cut from 1/4 inch and smaller rods of 52100 steel at the time of the change, but hardenable stainless steel, which the screw machines could not cut, was coming into the picture.

The change eliminated one operation—an end grind to size—because the abrasive wheels made a cut sufficiently accurate for the purpose, which the screw machines did not. On the other hand, three short operations—a barrel finishing step for deburring, a centerless "green grind," and a second barrel finishing step for radiusing the ends of the blanks—were added. The final center-type cylindrical (O.D.) grind remained unchanged.

Cast Iron Machine Base

These piece-parts were being milled top and bottom on a 30-horsepower vertical milling machine one at a time, with a total machining time of 19.6 minutes and a floor-to-floor time of 27.1 minutes. Tool replacement and tool resharpening costs represented 55 percent of the total cost.

When the part was switched to a 35-horsepower surface grinder on which two bases could be done simultaneously, machining time was cut in half, and floor-to-floor time by about 36 percent. That is about what could be expected from doing two at a time. However, the bigger saving was in tool replacement and tool resharpening, where the cost dropped by almost 83 percent. What actually happened was that replacement cost was cut in half, and resharpening cost was eliminated entirely.

However, the success of this first switch, plus the fact that the company had a number of similar "bread-and butter" jobs, justified the purchase of a still bigger 100-horsepower vertical-spindle surface grinder on which four bases could be machined at the same time. This time, machining time dropped to less than half that of the first grinder, 3.62 minutes, down about 82 percent from the time on the vertical mill, and floor-to-floor time went down to 11.23 minutes, down by 59 percent from the corresponding time on the milling machine. The machining time, it should be noted, was more than a full minute under what would be expected from a straight mathematical calculation. (If one at a time takes 19.6 minutes, four at a time would be expected to take 4.9 minutes each.)

From the difference in the reductions of machining time and floor-to-floor time, it is obvious that handling time for each base remained constant.

Tongues for Press Brake Dies

For another example, take the finishing of rough bars from which tongues for press brake dies are made, an operation involving the machining of two opposite sides of steel bars ranging from 1 × 3 inches to 1 × 6 inches in cross section, and either 10 feet, 6 inches or 12 feet, 6 inches in length.

When the finishing was done on planers, it took from 1 1/2 hours to about 2 hours to machine both sides of a bar. But when a vertical-spindle, reciprocating-table grinder was used, the corresponding times dropped to 15 minutes to 1/2 hour—and the operator still had time to dress the abrasive segments for a finishing pass which produced a better finish than was possible on the planers. The reduction in time released three planers for other work.

Crush Dressed Wheel Applications

Some possibilities of crush dressed wheels has been discussed in both chapters 5 and 6, relating to flat-surface grinding and to cylindrical grinding. Numerous makes of both surface and cylindrical grinders have been adapted for crush form grinding, and there is at least one cylindrical-type grinder built especially for this application. These machines produce formed parts at rates competitive with those of automatic lathes and screw machines, (Fig. 11–3), and with better dimensional and metallurgical characteristics than the turned parts.

Crush rolls are made very accurately, making it possible to hold the

Fig. 11-3. Grinding wheels with formed faces reemphasize a point made earlier: form grinding from the solid is one of the best possibilities for cutting costs and improving product quality through machining with abrasives. (*Hitchcock Publishing Co.*)

lateral locations of parts within 0.0002 inch, with general dimensional tolerances of about 0.0005 inch and surface finishes as good as 12 microinches. Depending on the form—deep forms and sharp corners reduce the number of parts per dress—one may expect to grind from 10 to 200 parts without redressing the wheel.

Parts with tolerances closer than 0.0002 inch or forms deeper than 1 inch can often be crush ground as a roughing operation and then finished with some other type of grinding.

Metallurgical Test Specimens. The production of metallurgical test specimens, sometimes called "coupons," is a routine operation in many steel or other metal-producing mills, in customers' plants, and in laboratories. The traditional way to make these has been to cut a 6 inch bar of material and then turn and polish a section in the center to a specified diameter. The customary time is 45 to 60 minutes. However, on a crush-form cylindrical grinder, it is one operation instead of two, completed from the solid bar in less than six minutes, a time saving of 85 to 90 percent. Tolerances are closer; surface finish is better, without discoloration or damage; and residual stresses are reduced. Specimens can be cut from any metal, regardless of its hardness or other qualities. (Fig. 11–4 A and B.)

Threaded Seals. In one plant, a threaded or labyrinth seal (Fig. 11–5) which was formerly thread-milled at a rate of 18 pieces per shift was a problem because the cleanup after milling took more time than the milling. The part required the forming of a number of 0.090 to 0.100-inch deep finished grooves. On a crush form grinder these could be done at a rate of 10 per hour, a fourfold increase in production.

Another fugitive from the thread mill was an angle seal, with grooves similar to the preceding part but cut on a 10° angle. The thread mill produced 18 pieces per shift, the crush form grinder, 40 per shift. This change is still a doubling of production or a halving of machining time, although it is not an improvement of the magnitude of that for the labyrinth seal.

Typewriter "Filter Shaft." This crush-form grinding operation approaches "big-chip" machining in terms of the size of chips produced, and it was credited with opening up a major manufacturing plant to the possibilities of crush-form machining.

The part in question is a typewriter part called a "filter shaft," a cylinder of carburized and hardened 1018 steel, 13 1/2 inches long and 5/16 inch in diameter, with two 3/32-inch ribs 180° apart running the length of the shaft. On the left end, ten different diameters had to be machined; on the right end, four. Every time a key is struck, the shaft makes a half turn; and when the machine is run from a tape, this could mean 1200 rpm for 2400 characters per minute. So the part has to be precision-finished.

Turning was tried, but the longitudinal ribs defeated the lathe tools, and neither production level nor part quality was satisfactory. But the switch to crush form grinding cut two complete steps from the operation, and engineers at the plant estimated that the cost savings were on the order of 25 to 35 percent, a conclusion that the reduction in the number of steps seems to confirm. Formerly, the operation involved these steps:

Cut to length (13 1/2 inches)
Turn ends
Heat treat
Anneal
Straighten
Grind

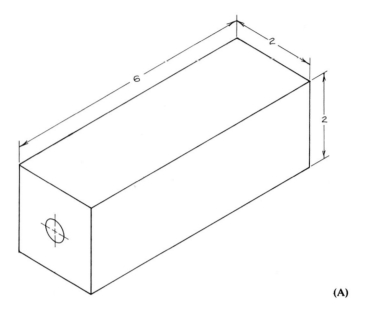

(A)

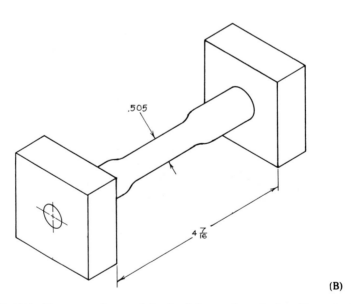

(B)

Figure 11-4(A)   The test specimen as it is after being cut from the bar. Figure 11-4(B) The specimen after it was ground in about 6 minutes. Turning and polishing took 45 to 60 minutes. (*Bendix Corp.*)

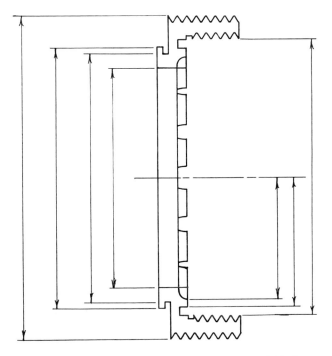

Fig. 11-5.   This labyrinth seal, formerly thread-milled at a rate of 18 pieces per shift and taking more time to clean up than to mill, is crush-ground at 10 pieces per hour. The grooves are 0.90 to 0.100 inch deep. (*Hitchcock Publishing Co.*)

For the crush-form grinding operation, the steps are:

Cut to length (14 1/2 inches)
Heat treat   (Turning and annealing eliminated.)
Straighten
Grind

Actually the parts are ground on two identical machines set up so that the operators are back to back; one machine grinds the diameters on the left end, and the other, the right end diameters. The extra inch of length of the shaft is to allow for chucking, but since both grinding operations are automatic—one a 22-second cycle and the other a 24-second cycle—there is time for an operator to snip off the ends of the shafts after they are finished. For practical purposes, the two cycles are of the same length.

Chips from this operation are reported to be 7/8 to 1 inch in length, 1/16 inch wide, and 0.002 inch thick, unusually large for grinding, but also an indicator of how fas the job got done. The operation involves an interrupted cut at the beginning of the cycle because of the longitudinal ribs on the shafts, a condition that is no problem in grinding but literally impossible in turning. Finally, a high-pressure jet of coolant at the back of the wheel away from the grinding area, which was mentioned earlier in connection with crush-form grinding, increases the intervals—and the number of parts that can be ground—between interruptions for wheel dressing.

Summary

The foregoing examples, as well as others that might be cited, indicate that many plants turn to some form of abrasive machining with wheels only because the traditional methods of cutter tool machining and finish grinding fall short in either level of production or quality of finished product. However, something should be said for recognizing the possibilities of using abrasive wheels more extensively.

One successful salesman for a grinding machine builder used to say that if he got on sufficiently good terms with a company to be allowed to walk through their machining departments and observe operations, he would have no trouble spotting operations where abrasive machining would be of benefit. His motto was, "If it has to be ground anyway, why not do the whole job with abrasives." It's worth considering.

COATED ABRASIVE BELTS

Much of what has been said about abrasive wheels applies with equal force to coated abrasive belts. Over the past few years, in fact, some makers of belts and builders of belt machines may well have been more active in promoting their products against milling and similar processes than were their counterparts making wheels and machines. And, as was noted in Chapter 8, both wheels and belts are definitely competitive with one another.

Proponents of abrasive belts maintain that their product is more uniform than are wheels. An earlier reference was made to the fact that of necessity the wheel-molding process resulted in totally random orientation of the grain in the wheel, while electrostatic deposition of the grain on adhesive-covered belt backing material provided optimum grain orientation. Belt thickness also tends to be quite uniform, within 0.001-inch variation over the 4-foot width of a jumbo roll of abrasive belt. And since most belts tend to be narrower than 4 feet—12 inches is a common width—the uniformity between belts is likely to be high.

The work material is usually cast iron or aluminum, although belt use on a number of steels and zirconium has been mentioned. Belt life ranges from 4 to 12 hours, depending in part on the work material and the severity of the application and probably to a degree on how well the machine is designed to spread the work-belt contact over the entire belt surface.

From published reports it would appear that there has been a growing movement toward the replacement (by the automotive industry as well as structural equipment builders and agricultural equipment manufacturers) of milling and wheel grinding machines by machines using coated abrasive belts. Whether this can be considered a trend is a matter of opinion, but some of the figures are impressive.

Automotive Applications

One high-production belt surface grinder is grinding cast-iron cylinder heads at a rate of 120 per hour, with 0.125 inch stock removed, 0.002-inch flatness, and 125-microinch surface finish.

An automotive valve body that is a combination of cast iron and aluminum is finish-ground, burr-free, at a rate of 250 per hour. Stock removal is 0.090 inch; flatness, 0.001 inch.

Grinding with belts on this machine leaves a short, noncontinuous scratch pattern which provides an excellent nonleak sealing surface for adjoining faces. One automaker is reported to be considering using the machine to make gasketless cylinder heads on a production basis.

This particular grinder (see Figure 5-20) uses a 30-inch wide belt traveling at 8000 sfpm, with grinding pressure in the range of 350 to 700 psi. Specific grinding energy (hp/cu in/min) is reported to be about 1.5, well down from a vertical surface grinder's 4.5, and competitive with milling. Of course, horsepower requirements, though important, are not the only consideration. The final consideration is metal-removal cost, and that was reported as 1 cent to 2 cents per cubic inch of metal removed.

Grinding can be done either wet or dry. Up to 96 percent of the cutting energy goes into the chips, which means that the workpiece remains cool and can be handled comfortably right after grinding. It also means that some means of cooling the chips is necessary—in this case by a curtain of water—because otherwise the chips might weld to each other and possibly to the machine and the piece-part.

Gray Iron Gear Case Casting

Another example involves the replacement of milling by a two-step rough-and-finish belt-grinding operation. The part consists of two halves of a gray iron gear case casting for an electric drive unit; the operations include the machining of the two mating surfaces of the halves, plus the mounting surface on the top half—three surfaces per unit.

The company had been machining these castings on a 30-horsepower milling machine, and the setup required elaborate fixturing to prevent distortion of the part by the considerable pressures involved. The milling cutters used 24 inserts at a cost of $1.79 each, for a total of $42.96; each set of inserts could complete 288 surfaces, or 96 units. Tool change time was 60 minutes. Actual milling time was 10 minutes per surface.

On the belt grinder, one belt costing $15.47 ground between 160 and 180 surfaces, or 53 to 60 units. Belt change time was 1 to 1 1/2 minutes, and actual grinding time was 58 seconds per surface.

These two operations were comparable in that both achieved the 250-microinch finish required to meet the basic specification that they had to be leakproof when bolted together with a gasket. However, the belt system was sufficiently better to justify the addition of a second flat platen grinder operation to improve the mating surfaces to 125 microinches and thus to improve the quality of the end product.

These data are summarized in the tabulation below. Tool change time has been converted to reflect per-surface times.

	MACHINING TIME	TOOL CHANGE TIME
Milling	10 minutes	12.5 seconds
Belt grinding	58 seconds	0.5 seconds
Time advantage	9 minutes	12.0 seconds
Percent improvement	90	96

At the time this was reported, the machine was handling the company's requirements for the gear cases with time to spare for finishing machine outlet boxes (cases and covers) and holding 0.0007- to 0.001-inch tolerances for UL standards.

Thin-walled Parts

Another type of application on which coated abrasive belts are attractive for cost and other reasons is the machining of either the sides or the ends of thin-walled parts. The broad area of coverage that is possible, together with the light pressure needed for cutting, often make belt machining either the favored method or the only possible method because it reduces the possibility of part distortion to a minimum.

Cast-Iron Speed-reducer Housings

This application was to machine the mounting surfaces on the speed reducers, and it had been in use successfully for about 2 years when it was reported. The company's management calculated that it would take two face mills and two operators per shift to equal the production of one 75-horsepower coated abrasive machining system and one operator per shift. Stock removal on this operation ranges from 0.125 to 0.190 inch in three to fives passes that are automatically controlled.

The short length of the cycle times makes fixturing very important. Over a dozen fixtures are used, to accommodate a wide variety of part sizes. Some are designed to hold one large part; others hold two smaller parts. Large castings are lifted into and out of the fixtures with a hoist; smaller parts are handled manually. In several cases it is possible to adapt the old manually-activated milling fixtures.

Clamping and release of the parts, as well as cycle time and other variables, are controlled by the operator from a master control panel on the machine. Changing fixtures on the belt machine took from 10 to 45 *minutes;* on the milling machine, the same operation took from 3 to 4 *hours* because of the multiple fixturing.

There is an unusual method for disposing of the swarf—chips and dust. The dry swarf is thrown against a curtain of coolant and when it hits that curtain, it becomes mixed with the fluid. The water-soluble coolant is passed through a cyclone filter which removes the swarf, and the coolant is then recirculated.

Theory of Abrasive Belt Cutting

One of the major suppliers of abrasives has come up with an interesting theoretical explanation of the cutting action of abrasive belts, one which to a considerable extent applies also to wheels.

On the surface of a 4- × 96-inch 50-grit belt, there are about 500,000 abrasive grains. The number would increase with smaller-size grains and/or with closed coating; it would decrease with larger-size grain and/or open coating. At an average belt speed, each of these grains will come into contact with the work surface somewhat more than 600 times per minute, thereby creating a theoretical stock-removal capacity of about 300 million extremely minute chips per minute. And even though it is assumed that in production, only one million of these grains actually remove a chip each, their volume will compare favorably with that of the single thick and continuous chip removed by a single-point tool.

With this in mind, one question that can come up is, will the job be better done with the removal of one continuous chip or with the removal of a multitude of smaller chips? If stock removal alone is the criterion, then the first might well be the choice. But where dimensional precision and quality of surface, plus flatness and/or parallelism or roundness, are involved, the second could well be preferable.

VIBRATORY FINISHING WITH ABRASIVE GRAIN

The abrasive grain operations as a group are almost entirely final finishing operations. Most of them are designed to produce either a good-looking finish, as, for example, on cooking ware, or to produce a surface of measured quality, smoothing out and softening the minute scratches remaining from prior operations. None removes much stock.

For these reasons, abrasive grain operations as a group are not really competitive with cutting tool operations, but blasting, for one, has been making inroads on acid pickling of steel sheets, rod, castings, and other forms. The problem with pickling has always been the disposal of the used pickling solution, which has not improved with current emphasis on the elimination of water pollution. The situation has probably become better with the substitution of nitric acid for sulfuric acid, but disposal is still a problem. In blasting, however, recirculation of the blasting media is part of the design of the system; the media is too valuable to be used once and discarded.

In the opinion of many experts in the finishing field, vibratory finishing offers probably the best possibilities for wider use, even though some of the processes displaced are also abrasive-using processes. Batch-type barrel finishing and vibratory finishing are slowed down because the machine must be stopped for unloading and reloading. Manual buffing and polishing with cloth wheels are hampered because the only way to increase production is to increase the number of operators together with the number of machines and the amount of space, though there are, of course, quite a number of high-production buffing units in operation, primarily in industries where appearance is the primary objective.

When it comes to high-production vibratory finishing, there are really two basic approaches. One is to have the finishing done in a long troughlike unit in which the piece-parts can be vibrated with media and compound. Parts and media are fed into the system at one end of the unit, which is sloped at such an angle that parts and media progress toward the discharge end, and when they reach that point, the parts pass across a vibrating screen to a tote box or conveyor, while the media is recycled to the loading end.

The other approach is to use a round unit in which parts and media circulate in a toroidal pattern around a doughnut-shaped tub. At the end of the run, a sort of dam called a swing blade—a wire mesh of appropriate size and shape—is lowered into the tub, the direction is reversed, and a gentler vibrating motion causes the parts to "walk" up the blade either onto a conveyor or into a tote box, while the media falls through the mesh for reuse.

Both systems probably do comparable jobs; their efficiency depends to a large degree on the selection of media and compound. However, in the important factor of floor space, it can be said that either takes less space than the number of individually operated floor stand polishers or buffers required for mass production, and that the round units probably need less space than the throughfeed types.

Booster Assembly Housing

This housing is a caplike piece of free-cutting brass about 2 inches in diameter and 1 1/2 inch in length, threaded inside and out, with several internal cavities, weighing about 8 ounces apiece. At a production rate of three million parts per year, the plant was employing 16 operators per shift for manual finishing.

After the parts leave the final machining operation, they must be thoroughly cleaned of all oil residue and other dirt, and all burrs and chips must be removed. With the previous manual buffing and cleaning, keeping up with the 20,000 per day production rate was a challenge. But with the installation of continuous cleaning and deburring with throughfeed vibratory finishing, three operators per shift can finish 20 parts per minute and on a two-shift basis can essentially keep up with the production level.

The vibratory is a long, neoprene-lined trough with a vibrating mechanism; parts and media move at a steady 40-minute pace from the loading end to the discharge end. Parts are loaded at three-second intervals at one end by an operator, picked up and transferred to a mesh conveyor belt by a second operator, and washed by a spray of cleaning compound that removes any loose material. After cleaning, a third operator carefully packs the parts in boxes for transfer to the assembly department. Obviously, it would not be a major change to make the loading automatic, one of the improvements that was being considered at the time the report was written.

Undersized media is screened out at the discharge end of the unit, and usable media is recycled for reuse.

Heads of Golf-Club Irons

In a matched set of golf clubs, the weights of the heads of the irons must be uniform within a "point," a point being about the weight of a dollar bill. This may help explain the relatively high classification and pay of operators who were finishing the heads by smoothing and polishing them on conventional bench-type belt grinders, and then finishing them by buffing. The only problem was that there was no way to increase production except by adding more belt grinders, more buffing machines, and more operators. And at the time of the report production volume had been going up by about 15 percent per year for 5 years—sufficient reason to check out different methods.

Vibratory finishing, one of those checked, looked good, except for the requirement that there be no nicking of the heads. Since the "hozel," or shaft end of the head, must be drilled and tapped, there was a built-in holding possibility. What evolved was a squirrel-cage type of holder on which 300 heads can be mounted at one time by loosely retained holding bolts with threads matching those of the hozels.

Loading and unloading are two-man jobs; one man holds a head against a bolt and the second tightens the bolt. Both loading and unloading can be easily accommodated within the finishing cycle, so that the vibrator runs practically full time.

The fixture is always suspended from an overhead crane which swings it between the vibratory and the load-unload area. In the actual finishing, the orbital motion of the media pushes the fixture around on a horizontal spindle, and media speed is sufficiently greater than that of the work to ensure finishing of the heads without loss of weight or change of shape.

Company estimates were that labor time had been cut at least in half, and that finishing capacity had been boosted by about 40 percent with no loss in the quality of the product. Other savings—floor space is one that probably occurred—were not mentioned, but the best outcome may well have been the capability of keeping up with considerably higher levels of production without significant increased costs.

SUMMARY

In the final analysis, comparisons between machining with abrasives and machining with cutting tools have to be considered in view of the fact that there is essentially one form of producing flat surfaces with cutting tools—milling; and one form for cylindrical shapes—turning. Furthermore, the choices are relatively simple, from among a small number of tool materials. (In the opinion of many, planing and shaping, also for producing flat surfaces, are relatively specialized.)

In machining with abrasives the choices are much more complex. For producing flat surfaces there are more than a half-dozen basic methods to begin with, each with some variations. Moreover, the methods and the techniques are different. Machining with the periphery of a wheel is a different operation from machining with the rim or flat of a wheel or with segments. And the choice of the best wheel can be complex, also. Finally, if the job is done with abrasive belts, it is still a different story.

So the replacement of a cutting tool machining operation with an abrasive machining operation is not simply a straight substitution. It is a matter of picking from the various abrasive possibilities the one that will do the best job. And that is not always easy, although it can be rewarding with respect to reducing costs. Of course, in any given shop or department one usually has to work with the available machines and processes. But when it comes to selecting replacement machines, as the foregoing reports emphasize, it is possible to reduce costs by informed use of the various processes for machining with abrasives.

12
The Bottom Line

A tenet of modern industrial management is that every operation ought to be done by whatever processes contribute most overall to profitable operation. It is difficult to quarrel with this as a theory. But a process change that benefits the overall company picture but increases costs in one department is not likely to be enthusiastically supported in that department. And the closer one gets to the actual operation, the greater, usually, is the resistance to change.

The following story may illustrate the point.

Overheating of a piece-part, commonly called burning, is generally an indication that the wheel is too hard or the infeed (downfeed) too severe. Burning usually results in a discoloration of the work surface. When this happens, the conventional wisdom is to change to a softer wheel in spite of the increased wheel cost and to reduce the infeed. This action may also reduce the production rate.

However, there is a well-documented point of view that if you go in the opposite direction—that is, if you increase the infeed with the same wheel—some interesting things happen. Dull grains are torn out of the wheel and replaced by new, sharp grains so that the wheel face remains sharp and free-cutting instead of dulling and glazing. The net result is no more burn. Finally, the production rate stays up.

But many operators don't like it; in fact, many regard it as a safety hazard because they are unsure of what the increased feed rate will do.

Suppose that on a particular operation an operator is getting burn on a given part with an infeed per pass of 0.020 inch. So somebody calls in a service engineer from either the wheel manufacturer or the machine builder. And suppose that either or both recommend doubling the infeed, with the same wheel, to 0.040 inch, and demonstrate to company management and the operator that at that infeed the wheel produces well and without burning the piece-part. Further, the wheel is rotating at a safe speed, which probably has remained unchanged.

But the minute the operator is alone, he probably cuts back the infeed to the 0.020 inch that caused the problem in the first place, and within a very short time the service engineer gets a call that the burning of the part is recurring. And the operator is very likely to tell the engineer, "You can feed it like that if you want to, but I won't."

The solution to *that* problem lies outside the scope of this book. However, the anecdote illustrates that there is frequently more to a change in method than the comparative efficiency of the two methods.

The availability of machines is another often crucial factor. It sometimes happens that an operation is done on an inefficient machine rather than to go through the hassle of justifying the need for a better machine. Or maybe the applicable budget will simply not stretch to provide for the purchase of the better machine.

Points of view among those concerned with machining probably range all the way from, "We grind only when nothing else will do the job" to "If I had my way we'd grind everything." Probably the majority tends toward the first viewpoint, but there is something to be said for the view that any part which must be ground at some stage in its processing ought to be checked out to see if it can be ground from the "solid" or the casting, or from whatever condition it is in when it comes to the machining stage.

SOME MISCONCEPTIONS ABOUT COSTS

Before getting into any discussions of the elements of machining cost and how they might be judged for either cutting tools or abrasives, it would be well to clear up a couple of misconceptions regarding surface finish and its costs.

Surface finish is a measurable element in the machining of any piece-parts, even though, as with automobile bumpers or frying pans where the only objective is good appearance, it may not be measured. And the measure of a surface finish, regardless of the way in which it may be expressed, is essentially an average of the depth of the scratches left by the preceding operation. After any of the preliminary or roughing operations, the scratches are relatively deep and the numbers consequently bigger. The concern at the beginning is that the scratches not be so deep that they cannot be removed in the subsequent machining. The final concern is to achieve whatever quality of surface finish that has been specified. The customary unit for expressing the quality is the average number of microinches of depth of the scratches, and it is generally thought that the better the finish (i.e., the shallower the scratches), the more expensive the finish.

For cutting tools the general opinion is correct. There is almost a straight-line relationship between the quality of the finish and its cost. But a different sort of relationship between quality of finish and its costs exists for machining with abrasives. Costs remain relatively level, usually, until the finish is in the vicinity of 12 to 16 microinches, maybe even somewhat better. Somewhere in that range they begin to rise; and in the very low range, under 16 microinches, costs probably rise more rapidly.

This is a very broad generalization, the essential point of which is that the relationship of cost and finish for abrasives is not the same as that for cutting tools. And the point at which the abrasive costs begin to rise is beyond the capability of any but the very fussiest of cutting-tool machining.

The situation in grinding costs can produce some awkward situations. There is a story, probably apocryphal, of an engineer with one of the major automobile manufacturers who was in charge of the honing of cylinders. When he found out that he could achieve a substantially better finish on the walls of the cylinders without increasing costs, he went ahead and specified it without considering the effect on oil consumption. As it turned out, there was no place on the cylinder walls for droplets of oil to accumulate, so the engine blocks with the improved finish on their cylinder walls used significantly more oil. Needless to say, the order was quickly rescinded.

Costs of Wheel Speed and Pressure

There is a long-standing consensus among research engineers and other similar personnel from the machine builders and the wheel manufacturers that most grinding machines are underpowered for most efficient use of the available wheels. This is probably true. Consequently, many machines on shop floors today are not capable of taking advantage of the wheels that are available. Many owners of such machines would do well to consider replacing the old machines with higher-powered grinders capable of higher pressure and higher speeds, so that the whole operation would become more efficient.

High speed—generally considered as anything over 6500 sfpm for vitrified wheels and 9500 for resinoid wheels—does put greater strain on any grinder, as does higher pressure. There is an extra safety hazard to high speed particularly; the wheel's safety guard must be both stronger and better designed to reduce hazard in case of wheel breakage. Modern machines generally conform to the highest standards. As an example, for quite a number of years it was considered that a round safety guard, closely hugging the wheel, was the best. Then it developed that in case of a breakage, the wheel fragments would speed around the inside of the guard and out without losing much velocity. Such guards are now designed with square corners which are much more effective in slowing down fragments.

Machines designed for high wheel speed and pressure must be more massive and rigid—and consequently more expensive—than those for low speed and pressure. But the extra mass brings with it another benefit— better dampening of vibration, which is a most desirable machine quality.

The cutting power of an abrasive wheel or belt, as has been pointed out earlier, depends on the speed of the tool and the pressure on the work for its cutting efficiency. Moreover, on applications such as foundry snagging, where the major objective is stock removal, both speed and pressure are above what they are in other applications.

A few years before and after 1970, there was a great deal of interest in increased speed as a means of increasing efficiency and cutting costs. The reasoning was that with increased speed, more cutting abrasive grains would be introduced into the wheel-work contact area, more stock would be removed faster, and overall costs of the operation would be reduced. And some extensive research was done concerning wheel speeds in the 18,000 to 20,000 sfpm and even higher range.

The essential conclusion of the research was that increased speed did increase the stock-removal capacity of the wheels, but not in a straight-line relationship. As speed increased, stock removal would also increase for a time; then there would be a pause or a plateau; and with further increases in speed there would be another increase in stock removal, then another plateau, and so on.

The move toward higher wheel speeds has decelerated in the past few years, partially because of the expense of the machines, partly because the lack of complete agreement as to the way to go, but probably mostly because of the increased hazards of the higher speeds. The research, it might be noted, was done on machines that were completely shielded and controlled from outside the room. And that, it goes without saying, is not a production setup.

If there were to be a general movement toward higher wheel speeds, there would have to be a whole new generation of machines, with substantially higher price tags.

The men who regard higher wheel speeds as not necessary generally lean toward greater pressure on the wheel or the belt as a solution. And at normal belt or wheel speeds, the number of abrasive grains to which the work surface is exposed is in the millions per minute, far and away beyond the number of cutting points that any cutting tool has available. Further, increased pressure is possible, without particular modifications, on a great many machines now in use.

For many years, the major obstacle to numerical control, or automation, or automatic operation of grinding machines was the problem posed by wheel wear, particularly the method of compensation for wear to maintain dimensional tolerances. This has now been dealt with, for there are machines available that will operate virtually without human intervention. It might be regarded as the final step of a progression that begins with, say, a two-worktable surface grinder set up so that one table can be unloaded and loaded while the piece-parts on the other table are being machined. True, it is only for use in very high production applications, but where it can be used it saves considerably in labor costs, to mention only one savings aspect.

Cost Considerations in Process Changes

Almost every plant of any size has some kind of organized procedure for justification of either methods or equipment changes, so there is no reason here to go into any detailed or extensive discussion of such procedures. The important point is that no change should be made on the basis of one factor alone, even though there may well be cases where one factor might be dominant.

Any such plan for change should include, beyond direct labor cost, analysis of at least the following: employee attitude, overhead, cost of materials, and tool cost—including the cost of maintenance and resharpening where applicable—equipment cost and power cost. There may be times where floor space required also becomes a factor.

Direct Labor Cost. Direct labor cost per unit or per 100 units is probably the easiest to figure. It is simply the number of parts machined per hour or per other unit of time, divided into the hourly rate, figured either with or without an amount for fringe benefits. The last element probably does not make any real difference, because it is a constant figure across the board.

Employee Attitude. While employee attitude may not be a cost figure, it is certainly a factor to consider in analyzing whether a change is actually beneficial or not, as was demonstrated by the anecdote about the operator who reduced the infeed to its former, and problem-causing, level just as soon as he had the opportunity. This is not the place for a full discussion of the point, but certainly if either an employee's earnings or his security is threatened by a change, he is not going to support it wholeheartedly. There is a commonly held opinion that many workers can make any comparison test come out the way they want it to, which may be why many salesmen of coolants and wheels or belts like to be on good terms with the operators.

Overhead Costs. Overhead is likely to be an amorphous concept to many people at the foreman or floor process engineer level. In many companies it is a figure furnished by the accounting section as an average or constant to be included in calculating the benefits of a process or machine change. And if that is plant policy, then it is what should be done.

But as every foreman knows, all operations are not equal; some run smoothly day after day, while others demand a lot of time and attention. The ones that fall into the latter category should be among those to be examined to see whether they could be done by some other method.

Material Cost. In previous chapters the possibility was mentioned that a change to some form of machining with abrasives may permit a redesign of a part to save material. Also mentioned was the fact that improvement in the ability of previous operations to provide a part for machining which is closer to finished dimensions is an encouragement to change from cutting tool to abrasive machining. The flip side of this, also stressed, is that a cutting tool requires a certain minimum amount of stock to be effective, whereas a grinding wheel or a belt will function efficiently with the least amount of stock the previous operations will provide. If the latter is the case, then there ought to be a notation somewhere of the amount of material saved. This is a factor which is becoming more important every month.

Tool Cost. The most important thing to remember in comparing tool costs between the two principal methods of machining is that a cutting tool must be periodically removed from the machine, transported somewhere else, resharpened, returned, and eventually reinstalled on the machine. On the contrary, a grinding wheel, once mounted on a machine, will frequently stay there until it is worn out, needing only periodic dressings which are done by the operator. Perhaps the major caution about that operation is to ensure that the operator does not overdress the wheel, that is, that he does not remove more grain than is needed to restore the surface to efficient cutting, and that he does not dress the wheel more frequently than is needed. It is obvious that mounting a grinding wheel on a machine is practically always more time-consuming than is mounting a cutting tool. There is no dressing for abrasive belts, but the belt must be replaced on some schedule, frequently on the order of once or twice per shift. This was once a rather cumbersome operation, particularly with big belts, but the designers have come up with devices that generally keep the downtime for changing belts to 2 or 3 minutes.

In short, the original cost of any tool is not the only cost to be considered. If it were, diamond and CBN wheels would never have been used.

One final point must be brought up, however: the matter of averaging the cost of resharpening tools. An average may serve a number of useful purposes, but accurate costs on a single group of tools is not one of them. The average lumps together both tools that need only a 5-minute touchup along with those whose cutting edges have become very dull, perhaps chipped, and need perhaps an hour to recondition. Neither one is a representative time. It is not always easy to get a reading on the customary time for resharpening a particular tool or group of tools, but it is worth some effort on most production jobs to have something more than a guess.

Equipment Costs. In every process change that involves the purchase of new equipment, the question that must be answered is, "Can we afford it?" In a small shop the owner is likely to make the decision based on some mental calculations from his knowledge of every facet of the business. In most larger companies the procedure is more elaborate, and the decision of affording the expenditure comes down to a calculation of the pay-out period—the time that it will take, from estimated savings, to recoup the expenditure.

There might be one variation on this. In the life of every machine there comes a time when it must be replaced. In many plants, judging from the average life of machine tools nationwide, the "must" replacement point is some time after the "should" replacement point. And finally, there is always the question of when an old machine is sufficiently obsolete to justify its replacement.

Power. There seems to be little doubt that a lathe tool or a milling cutter is more efficient than a grinding wheel in removing excess stock from a piece-part and that an abrasive belt lies somewhere in between the two. But neither abrasive process requires the amount of excess stock that is a necessity for turning or milling. So, part of the analysis concerning power may boil down to a choice between a more-effective method that also must have more stock to remove and a less-effective method that does not require as much excess material.

Numerical Control of Grinding. Our discussion so far has primarily been concerned with abrasive machining as an alternative to cutting-tool machining as processes only, without considering particularly how either might be designed for control by instruments or tape or other replacements for human hands or minds. And the only references to grinding machines for other than the straight cycle of load, process, and unload have been mentions of duplex surface grinders with more than one table, or belt surface grinders with several work stations, or incorporation of dressing into grinding cycles.

For many years the obstacle to nonhuman control of a grinding operation was the unpredictable wear of the wheels, which led to out-of-tolerance parts. But new developments in wheels have made wheel wear much more predictable, enough so that constant manual checking or observation is no longer needed. The elements of the grinding cycle, including dressing, can be programmed to proceed with only occasional monitoring.

Furthermore, some of today's grinders—centerless and double-disc grinders are two probable types—can be incorporated into production transfer lines down which parts move from operation to operation automatically. And of course there are special grinders that can perform operations on several faces of a piece-part in sequence; but the key to such operations is in the handling and conveying elements, not in the basic grinding process. And the machines themselves are usually elements in the entire system designed to finish one complex part.

Possibilities for Using Abrasives

This book has been based on the assumption that more knowledge of what the various abrasive processes can do, particularly in comparison with cutting tools, would aid in a wider use of abrasives and more profit to companies that increase their usage. Not everyone can be as sensitive to such profitable changes in methods as the salesman who was mentioned earlier, who needed only to tour a plant to spot them; but the knack is not difficult to acquire, once one has some hints about what to look for. Here are some of the indicators of possibilities for profitable switches to machining with abrasives.

Parts Already Partially Ground. Any part that for reasons of dimensional tolerances or finish specifications must be finished with abrasives ought to be checked to see whether all the machining could not be done with abrasives, possibly even on one machine, thereby reducing or eliminating transportation. As was pointed out earlier, this may involve some examination of the processes upstream from machining to see whether excess stock can be eliminated or the part redesigned to make it more suitable for grinding.

Any part being milled and ground can possibly be ground only, on a surface grinder. Any part being turned and later ground can possibly be ground only, on a suitable type of cylindrical grinder. Parts being buffed or polished one at a time on floorstand belt grinders or similar buffing machines can probably be finished less expensively in a vibratory finisher.

Large Areas for Machining. Any flat-surfaced part which has a comparatively large area to be machined warrants examination to see whether the area could be reduced to make the part more feasible for grinding. The point about the adaptability of abrasive wheels or belts to interrupted surfaces has been stressed mostly with respect to flat surfaces; it applies equally to shafts. And whether the interruption is a longitudinal rib or a slot doesn't particularly matter. Such features make turning difficult if not impossible; they are no real problem with either a belt or a wheel.

Manual Finishing. Any parts being manually finished one at a time ought to be looked at, regardless of the method of finishing. For some products there is a good selling point in being able to say that they are hand-finished; it might be possible to provide an equal or even better product—certainly more uniform—by finishing in a through-feed vibrator or mill.

Forms. In machining, a "form"—a rather inclusive term—is essentially any part that is other than a flat surface or a shaft with just one diameter, and the possibilities of forming and finishing such parts from a blank or from the solid appear to be endless. This includes parts that are contoured or radiused or have multiple diameters or annular grooves or slots. It also includes parts with multiple slots.

Simple radii, either convex or concave, are readily dressed on a wheel's periphery with a radius dresser.

Other possibilities are use of either a formed diamond dresser block or a crush dresser, although neither is inexpensive and both require a fair level of production to justify their use. The dresser block is a base machined to the form desired in the finished part and then plated with diamond abrasive. The crush dresser also has the form of the finished part. The diamond dresser cuts the abrasive grain in the wheel and thereby produces a wheel cutting surface with more sheared-off grain and less sharp grain, making for a smooth finish but a lower cutting rate. Crushing leaves the wheel face with jagged, aggressive grain that cuts faster but leaves a somewhat rougher finish. Both have considerable possibilities, although crush form grinding has been more aggressively promoted over the past decade or so.

Coated Abrasive Belts. The capabilities of coated abrasive belt machines appear to be expanding every year, to the point where belt-type grinders are currently competing successfully with milling machines and lathes, as well as with bonded abrasive wheel surface and cylindrical grinders. Belts have no competition in terms of the area that can be covered per pass across a piece-part; they routinely sand 4-foot-wide sheets of plywood and steel sheet, and they can cover wider sheets with appropriate splicing, although the cost rises rather rapidly. The difficulties of making cloth backings flexible enough to go around the contact rolls on the machines but still strong enough to resist tearing under pressure held back the stock-removal development of belts for years, but within the past decade, suitable backings for such purposes have been developed and the obstacle, for practical purposes, has been bypassed.

There is probably an area of close dimensional tolerances in which coated abrasive belts may never compete with grinding wheels, but the area appears to be narrowing steadily. Belts probably qualify as stiff competition for wheels in practically all the areas listed earlier, form grinding excepted.

SUMMARY

The grinding wheel and the coated abrasive belt will not—maybe never will—oust the milling machine and the lathe from all machine shops for a long time to come. But with developments made over the past quarter century, there is reason to believe that the millions of little abrasive grains will someday be doing more of the work now routinely assigned to milling cutters and turning tools. Time and the increasingly finer dimensional tolerances and surface-finish requirements are moving industry out of the area of the cutting tool and into the area of the abrasive grain.

Someone has said that World War I was a war of thousandths (of an inch) as tolerances within the range of cutting tools and the primitive grinding machines and wheels then available. World War II required tenths (ten-thousandths of an inch), definitely in the abrasive range; and the Space Age has already brought us to the millionths range.

At such levels there are few alternatives to machining with abrasives.

Glossary

The following words or terms are commonly used in connection with the various material-removing and finishing processes discussed collectively in this book as machining with abrasives. Some are defined in the text but included here for completeness. Some may be mentioned, but need elaboration. Many have special meanings when used in connection with abrasives.

Abrasive. Any substance, primarily nonmetallic, used in small pieces to remove material by cutting. Individual abrasive-using processes include grinding, lapping, polishing, snagging, and others. Abrasives do not remove material by friction or heating. The major abrasive materials are silicon carbide, aluminum oxide, natural and manufactured diamond, and cubic boron nitride.

Abrasive Machining. A term introduced around 1960 to denote any process using abrasives to remove substantial amounts of material to close dimensional or surface-finish tolerances. It emphasizes the stock-removal capability of abrasives.

Aluminum Oxide. A hard abrasive substance made by fusing natural bauxite or alumina in an electric furnace. The most-used abrasive for machining steels.

Arbor. The spindle of a grinding machine, on which a wheel is mounted.

Arbor Hole. The hole in the grinding wheel which goes onto the arbor.

Arc of Contact. The portion of the circumference (or periphery or outside diameter) of a grinding wheel or belt in contact with the work being ground.

Area of Contact. The total area of a wheel or belt in touch with the work being ground. The area is typically largest in surface grinding with the flat rim of a grinding wheel or with the flat side of an abrasive segment and least in the grinding of steel balls.

Balance (static). A grinding wheel is in static balance when it remains at rest in any position on a frictionless horizontal arbor. Most grinding wheels are so tested during manufacture.

Barrel Finishing. A process for finishing workpieces in which the piece-parts, abrasive grain, and liquid are processed in a closed octagonal or hexagonal barrel rotating on its long axis. The abrasive action takes place as the mass tumbles over and over during the rotation. The older name for the process is tumbling.

Bauxite. A claylike substance high in aluminum oxide content. The principal deposits of this material in the United States are in Arkansas. The term is derived from the name of the French city Baux, where the substance was first discovered.

Bench Grinder. A small utility grinder usually mounted on a work table or bench.

Blending. The process of smoothing out rough areas on a workpiece to ensure that their entire surface has the same plane or roundness and/or the same surface finish.

Blotter. A disc of compressible material used to cushion the contact between the sides of a grinding wheel and the flanges between which it is mounted. It reduces slippage also. Maximum safe operating speed, original wheel size, and wheel formulation may be printed on it.

Bond. The material in a grinding wheel which holds the abrasive grains together and supports them while they cut. There are four common bonds for aluminum oxide or silicon carbide wheels: vitrified (inorganic) and resinoid, rubber and shellac (organic), and three for diamond or cubic boron nitride wheels: vitrified, resin, and metal.

Burning (or workpieces). A change in the structure of work being ground caused by the heat of grinding. Burning usually shows as discoloration or softening of the work.

Burnishing. A glazed surface finish usually resulting from using a dull or loaded (with work material) grinding wheel.

Burr. A turned-over edge of material resulting from punching a sheet, from drilling, or from some other machining process. Removal of burrs is usually called deburring.

Bushing. A soft metal like lead, babbitt, or aluminum used to line the arbor holes of some grinding wheels. Also a removable ring, usually steel, or adapt a grinding wheel to a smaller spindle.

Centerless Grinding. Grinding the outside diameter of a workpiece supported between two abrasive wheels on a work rest rather than on centers (see below). One wheel machines the work; the other, rotating at a slower speed, keeps the work from spinning.

Centers. Conical steel pins upon which work is supported, centered, and rotated during center-type cylindrical grinding or during turning.

Centrifugal Separator. A machine which cleans coolant or grinding fluid by mechanical rotation to remove bits of used-up abrasive or work material.

Chatter, or Chatter Marks. An undesirable and repetitive pattern created on the surface of a workpiece, usually at regular intervals, which is usually caused by some out-of-round of out-of-balance condition in the machine. Chatter occurs in both grinding wheel and abrasive belt machines.

Closed Coat. A descriptive term for any coated abrasive tool (see below) whose surface is virtually covered with abrasive grain. (See also *Open Coat*).

Coated Abrasive. Cloth or paper with abrasive grain glued or otherwise adhered to the surface.

Cone Wheel. A small bonded abrasive wheel mounted on a pin or mandrel, typically cone- or bullet-shaped, or a similarly shaped wheel made of a coated abrasive strip wrapped around a mandrel. Both are used primarily on portable grinders.

Contact Wheel or Roll. Usually a hard rubber "tire" mounted on a metallic center, which backs up a coated abrasive belt at its point of contact with the workpiece. The surface is divided into alternating grooves or slots and lands, which help determine the grinding action of the belt. Some steel contact wheels are used. Contact wheels are usually narrower than contact rolls, but both are usually collectively called contact wheels.

Coolant. The traditional shop name for any liquid used to cool workpieces during machining. Coolants control the temperature of the workpiece during machining, may also lubricate the wheel or belt interface with the workpiece, and help remove bits of used-up abrasive and minute chips of the work material.

Corner Wear. The rounding of the corner of an abrasive wheel from wear of the abrasive grain. The corner of the wheel is the intersection of the periphery and the side of the wheel. Corner wear is usually a problem in form grinding.

Corundum. A natural abrasive of the aluminum oxide type. Rarely if ever used today.

Creep Feed Grinding. A technique of plunge grinding with a specially designed surface grinding machine involving very low table travel speed, with the total amount of stock to be removed from the workpiece taken off in one or at most two passes instead of in numerous lighter passes as in conventional surface grinding.

Critical Speed. The speed at which a cone wheel or other wheel mounted on a mandrel begins to vibrate excessively, and beyond which further operation of the wheel would be hazardous.

Crush Dressing or Crush Forming. The process by which an abrasive wheel's periphery is formed by forcing it against a steel roll that has the contour desired in the finished part. Most effective on vitrified-bonded wheels.

Cubic Boron Nitride (CBN). An abrasive next to the diamond in hardness, manufactured by a process similar to that for making diamond, and priced like diamond. Extremely effective on hard steels.

Cup Wheel. A grinding wheel shaped like a cup or bowl, designed so that grinding takes place on the rim or wall of the wheel rather than on its periphery. A straight-sided cup is called a type 6 wheel; a flaring cup, a type 11.

Cutting off Wheel. A thin abrasive wheel, usually resinoid- or rubber-bonded, for parting bar stock, tubing, and like materials, or for slotting. Also called a cutoff wheel.

Cutting Rate. The amount of material removed by a grinding wheel or an abrasive belt per unit of time, usually expressed as cubic inches per minute.

Cutting Surface. The part of the wheel surface which grinds the workpiece, either periphery or side or rim, depending on the wheel. Also called the grinding face.

Cylinder Wheel. A grinding wheel with a large hole relative to its diameter, usually several inches thick, for grinding on the rim of the wheel. Also called a type 2 wheel. Dimensions are given as diameter, thickness, and rim width (rather than hole size).

Cylindrical Grinding. Grinding the outside surface of a cylindrical part mounted on centers. May include, but is usually distinguished from, centerless grinding, which is the grinding of cylindrical parts *not* mounted on centers.

Deburring. The removal of turned-over edges left from previous machining. See *burr*.

Diamond Wheel. An abrasive wheel made with a layer of crushed diamond, either natural or manufactured, around a core.

Disc Grinding. Abrasive cutting or machining on a machine on which one or two abrasive discs are mounted either horizontally or vertically. Discs do not have a center arbor hole, but they may have a pattern of mounting holes over the flat grinding face.

Dish Wheel. A grinding wheel shaped like a flat dish, commonly used for sharpening cutters. A standard type 12 wheel.

Dressers. Any tools used for reconditioning a grinding wheel by removing the outer layer of dull abrasive grain and bits of foreign material.

Dressing. An operation either to renew and sharpen the grinding face of an abrasive wheel or to change the contour of the face.

Durometer. A number expressing the hardness of rubber. The higher the number, the harder the rubber.

Emery. A natural abrasive containing some aluminum oxide. Rarely used today, but the name survives as a common but incorrect name for abrasive wheels or coated abrasive cloth. "Emery cloth" is now used only in home workshops.

External Grinding. Grinding the outside diameter of a workpiece—cylindrical or centerless grinding—as distinguished from internal grinding.

Face. The part of an abrasive wheel that does the grinding.

Face Grinding. Grinding a plane surface at right angles to the grinder's magnetic chuck or work table. The spindle holding the wheel is horizontal.

Feed, Cross. In surface grinding, the increment of horizontal movement of the wheel either toward or away from the operator. The distance that the wheel is moved across the work between passes.

Feed, Down. The distance that the abrasive wheel is moved into the workpiece after pass across the surface being ground. In cylindrical grinding, called *infeed*.

Feed Lines. A pattern of parallel (surface grinding) or spiral (cylindrical) lines on the workpiece, usually indicating that the outside diameter of the wheel is slightly concave. Usually corrected by re-dressing the wheel face and slightly rounding the corners.

Filling. In coated abrasives, the clogging of the abrasive surface with bits of work material and dirt. Similar to the loading of an abrasive wheel. Can be reduced by changing to an open coat belt or by using a grease to fill the space between the abrasive grains.

Fin. Any thin projection on a casting.

Finish. Surface quality or appearance. Quality is measured in terms of the average variation from a mean or average of the scratches left by a cutting tool, grinding wheel, or abrasive belt, measured in millionths of an inch (microinches). The method of calculation varies. One such is the root mean square (rms); another is the arithmetical average (aa); and a third is the centerline average (cla). The rating of the surface often indicates the method of calculation, as 16 rms or 24 aa.

Finishing. The final or sometimes a semifinal machining operation to achieve the dimensional accuracy, the geometry, and the surface finish required in a workpiece. Minimal stock removal.

Firing. A term applied both to the 2000+ °F temperatures needed for vitrifying clay-bonded (vitrified) abrasive wheels and sometimes to the under 500°F temperatures for hardening the thermosetting plastics of resinoid- and rubber-bonded wheels.

Flanges. The circular metal plates on either side of a wheel by which the wheel is driven.

Flaring Cup. A rim-type grinding wheel whose diameter at the grinding face is greater than the diameter at the back. The wheel wall is also thicker at the back than it is at the grinding face. A standard type 11.

Floorstand Grinder. In abrasive wheels, a two-wheel bench grinder mounted on a waist-high pedestal. A coated abrasive belt grinder is similar but has the extra back-stand idler pulley for cooler belt action.

Form Grinding. Practically any grinding that produces something other than a flat, plane surface or a one-diameter cylinder.

Free Abrasive Machining. A loose-grain abrasive operation resembling lapping, but using coarser abrasive grain with harder upper and lower plates on the machine than are used for lapping. It is claimed that this operation removes greater amounts of stock than does lapping.

Freehand Grinding. Any grinding in which the work is held against the wheel or belt by hand. Also called *offhand grinding.*

G- Ratio. A measure of the relative wearing away of the work and the abrasive wheel in grinding. On soft materials the ratio will be high, indicating that the volume of stock removed is much greater than the volume of abrasive lost. On very hard materials with conventional abrasives the ratio may approach 1 to 1.

Glazing. A wheel-grinding face condition that is created when the abrasive grains become dull without being pulled out of the wheel. This usually indicates that the wheel is too hard for the application.

Grade. The strength of hold of the bond on abrasive particles, commonly referred to as the wheel's hardness. In a standard marking, grade is indicated by a letter immediately after the number indicating grain size, ranging in theory from A (very soft) to Z (very hard); but in practice, the softest wheels are probably E grade and the hardest, something like ZZZ grade.

Grain. Abrasive particles classified into predetermined sizes.

Grain Size. The standard nominal size of abrasive particles, originally based on the number of grains per linear inch. The size numbers still reflect this aspect: the larger the number, the smaller the grain. Specifications for all nominal sizes allow for a small percentage of grain larger than nominal, and for a somewhat larger percentage of grain smaller than nominal.

Grain Spacing. The relative density of the abrasive particles in a grinding wheel, termed "structure," and indicated by a numeral midway in the standard marking. The larger the number, the more open the wheel structure. If the structure is considered "standard," then the number may be omitted.

Grinding. The removal of material by any bonded or coated abrasive tool.

Grinding Action. Refers usually to the cutting ability of, and the quality of the finish produced by, any bonded or coated abrasive tool.

Grinding Fluid. Any fluid used either to control work temperature, to improve cutting action, to improve finish, or to remove bits of work material and abrasive.

Grinding Machine. A broad term which includes all machines on which bonded or coated abrasives are used. It is not so often used for machines using loose abrasive grain for finishing.

Grinding Ratio. Same as *G ratio*.

Grit Size. An alternate term for grain size; quite possibly more frequently used.

Hard Wheel. A term used by proponents of coated abrasives to describe a grinding wheel, without any reference to the grade of the wheel. Also, of course, a grinding wheel whose grade is R (approximately) or harder.

Honing. Once applied to virtually any fine finishing abrasive process, honing is now restricted to abrasive machining typically performed on internal cylindrical surfaces, employing bonded abrasive sticks reciprocated in a special holder to remove stock and improve surface accuracy.

Idler Pulley. The "other pulley(s)" with a contact wheel in coated abrasive belt machining system. Either the contact wheel or one of these pulleys may be power-driven.

I. D. Grinding. Another term for *internal grinding*.

Infeed. In cylindrical grinding, the distance that the wheel or belt is moved into the piece-part after each traverse, or at intervals.

Internal Grinding. Grinding the interior surface of any hole in a workpiece with a rotating abrasive wheel rather than, say, reciprocating sticks as in honing.

Jumbo Roll. A roll of coated abrasive as it comes off the final making machine. The material is just over 4 feet wide and usually about 50 yards long.

Land. In machining, a raised area, usually flat, lying between two grooves or slots. The grooves are generally parallel. (See Figs. 6–17 and 6–18.)

Lapping. A flat finishing operation employing loose abrasive grain. Sometimes used loosely to describe cylindrical finishing operations.

Loading. The filling of the pores of a grinding wheel with bits of the material being ground. Similar to "filling" of coated abrasives.

Lubricant. A liquid, usually an unmixed oil, whose primary purposes are to improve the surface finish of the work and to improve the cutting action of a wheel or belt. Good lubricants are rarely efficient at removing swarf resulting from grinding. (See also *coolants, grinding fluids.*)

Lubricity. The ability of a substance to form a lubricating film between moving surfaces, particularly under heavy pressure.

Machining. Cutting away excess material to form a finished part. Once considered as being done by cutting tools only, but now a process done with abrasives as well.

Mass Finishing. A collective term applied to several loose-grain finishing operations—for example, barrel finishing, vibratory finishing, spindle finishing—in which the workpieces or piece-parts are finished in batches, or continuously, rather than one at a time.

Media. A term (singular) to describe the loose abrasive used in mass finishing.

Mesh Number. Another term like grain size or grit size to describe the size of abrasive grain. It is derived from the number of openings per linear inch in the control sieving screens. Used more frequently for diamond and CBN grain, or for coated abrasives.

Microinch. One-millionth of an inch.

Mounted Wheels, Points. Small bonded abrasive shapes or wheels permanently attached to steel spindles or mandrels.

O.D. Grinding. A shop term for center-type cylindrical grinding.

Offhand Grinding. Grinding with an abrasive wheel or belt workpieces held in the operator's hand. Same as *freehand grinding.*

Open Coat. A term to describe coated abrasives whose surface is 50 to 70 percent covered with abrasive. Open coating reduces belt or disc filling or loading. (See *closed coat.*)

Operating Speed. The speed of an abrasive wheel or belt, expressed either as revolutions per minute (rpm) or surface feet per minute (sfpm). Rpm is the term usually applied to the operating speed of machine spindles or arbors, although it appears on many grinding wheels. Sfpm reflects the distance traveled per minute by a point on the surface of the wheel or belt, expressed in feet per minute.

Organic Bond. A wheel bond made of organic materials such as thermosetting synthetic resins, rubber, or shellac, which are fired (or baked or cured) at temperatures under 500°F.

Peripheral Speed. The speed at which any point or particle on the face of an abrasive wheel or belt travels when the wheel or belt is revolved at production levels, usually expressed in surface feet per minute (sfpm).

Piece-Part. A workpiece. Single unit of product in process. Usually a shop term.

Platen Grinder. A coated abrasive belt grinder equipped with a drive pulley and an idler pulley in a vertical line above. Between the two and facing the operator is a flat plate (the platen) against which the work is held and also supported by a work rest beneath. At the point of contact, the belt is traveling downward. Once primarily a tool for home workshop use for offhand grinding, the platen grinder has been redesigned in larger units for industrial use.

Plunge Grinding. Any grinding with either a wheel or a belt in which the abrasive element is fed straight into the workpiece, without any traverse (left-right-left movement) on the part of either. In this mode the wheel or the belt must be wider than the workpiece.

Polishing. Producing a high finish on piece-parts by applying them to a fine-grit abrasive belt or to a cloth wheel whose periphery is charged with fine abrasive in a liquid carrier. Polishing improves appearance but does not necessarily result in a better-quality surface finish. Polishing is frequently followed by buffing, a similar cloth-wheel operation with even finer abrasives.

Portable Grinder. Any grinder designed to be carried to the work.

Precision Work. Work required to be close to perfect in dimensions, surface finish, and other qualities. Dimensions probably to tolerances of 10 "tenths" (0.001 inch) or better, and surface finishes on the order of 32 microinches. Varies, of course, with general level of work in the shop.

Production. Department or section whose purpose is to turn out large quantities of whatever product the organization produces, usually with single-purpose machines by semiskilled operators, as distinguished from toolroom operations which involve small quantities or single parts processed by tradesmen or journeymen on multipurpose machines.

Profilometer. An instrument for measuring the degree of surface roughness in microinches.

Recessed Wheels. Grinding wheels with a centered circular area on one or both sides of the wheel which is not as thick as the measurement at the outside diameter of the wheel. With special flanges, such wheels can be used on grinding machines with short spindles to give the grinder greater capacity.

Reinforced Resinoid Bond. A synthetic resin (resinoid) bond with layers of material molded in to increase the ability of thin wheels to hold together at high speeds.

Resinoid Bond. A bonding material described commercially as a thermosetting synthetic resin. Its standard marking symbol is B, derived from Bakelite, one of the earliest such bonds used.

Rest. Any machine element which supports a piece-part from underneath. Rests can be found on bench grinders, snagging grinders, platen grinders, and centerless grinders. Steadyrests are used on center-type cylindrical grinders to support long, thin shafts that would otherwise tend to bow under the infeed pressure of the grinding wheel or belt.

rpm. Revolutions per minute.

Rubber Bond. A bonding material the principal of which ingredient is natural or synthetic rubber. Its standard marking symbol is R. It is used for making very thin cutoff wheels.

Segments. Shaped bonded abrasive sections designed to be fitted into a holder to form either a continuous or an interrupted flat grinding surface.

sfpm. Surface feet per minute, a measure of wheel speed.

Silicon Carbide. An abrasive produced by the reaction under heat of coke and silica sand in a resistance-type electric furnance. Used mostly for nonferrous metals and nonmetallic materials.

Snagging. Cleaning castings by rapid removal of excess stock, on swing frame or floorstand grinders with coarse-grit resinoid wheels. Heavy stock removal with minimal regard for finish.

Star Dresser. A hand tool equipped with rotatable star-shaped cutters, for rough dressing or large, coarse-grit wheels.

Steadyrest. See *rest*

Stock. Excess piece-part material to be cut away during machining.

Straight Wheel. A grinding wheel with three dimensions—diameter, thickness, and hole—which is not recessed, grooved, beveled, or tapered. A type 1 grinding wheel. Most wheels in use are straight wheels.

Structure. A standard wheel marking term generally describing the porosity of a grinding wheel, expressed by a digit midway in the marking symbol, as A80–K8–V20. The 8 indicates that the wheel is of medium structure. See also *grain spacing*.

Superfinishing. A finishing operation like honing, but with the shaped bonded abrasive sticks on the outside of a cylinder being forced inward.

Surface Grinding. Grinding a plane surface which either is parallel or has been set up to be parallel with the grinder's work table or magnetic chuck. The ground surface is usually flat, but may be contoured, as viewed from the end.

Swarf. The waste from a grinding operation, made up of bits of the work material, particles of used-up abrasive, coolant, dirt, and possibly machine oil. Swarf is usually circulated to the coolant tank with the coolant, and it must be periodically cleaned out.

Truing. A dressing-type operation whose object is to make the entire grinding face of an abrasive wheel equidistant from the center of the spindle on which the wheel is mounted.

Unit Horsepower (sometimes specific horsepower). The amount of power needed to remove 1 cubic inch of a given material in some unit of time, usually a minute.

Work Speed. In cylindrical, centerless, and internal grinding, the rate at which the piece-part revolves, measured in rpm or sfpm. In surface grinding, the rate of table traverse (left to right and back) measured in feet per minute.

Index

Index

Abrasive action, 34, 61, 86–96
Abrasive belt grinder heads, 109–110
Abrasive cutoff
    types of machines, 191–193
    wheels, 83–89, 187–188
Abrasive depth of cut, 252
Abrasive finishing machines, 196–198,
        203–206, 215–234
Abrasive grain, 37–51, 214, 217
    aluminum oxide, silicon carbide, 37,
        39–49
    crystalline structure, 48
    diamond, cubic boron nitride, 38,
        49-51, 74
    friability, 39, 45–46
    hardness, 4, 48
    shape, 48–49
    specific gravity, 49
Abrasive powders, 47
Abrasive wheel dressers, 92–96
Abrasive wheels (*See* Grinding wheels)
Abrasives
    examples of wider use, 283–285
    selection, 6, 56–57
    types of,
        aluminum oxide, 45–46
        for coated abrasives, 103–104
        silicon carbide, 41, 56–57
Accessories for surface grinding, 135–136
Acheson, Dr. Edward G., 39, 48
Adhesives for coated abrasives, 9, 105–106
Aluminum oxide, 3, 41
    for coated abrasives, 103, 104
    development, 42

    grain types, 45–46
    hardness, 4, 48
    manufacture, 42–44
    for non-ferrous metals, non-metallics, 5
    for steels, 5
Area of wheel-work contact, 35, 61–62
    belts vs. wheels, 118–119
    in I.D. and O.D. grinding, 174
Automation of abrasive machining, 219–220,
        272, 280, 282–283

Backings for coated abrasive belts, 10, 36,
        104–105, 211
Backstand idler, belt machines, 110, 114
Balancing of bonded abrasive wheels, 69, 70
Barrel finishing, 31, 214–218. (*See also*
        Tumbling)
    advantages, limitations, 215–216
    intermittent action, 215
    process, 218
    separation of media, parts, 217
Bauxite, 42, 43
Belts, coated abrasive, 98, 99, 109, 117, 119,
        209–213
    changing, 101, 211
    costs, 209, 211
    cutting rates, 100, 101
    speed, 119
    vs abrasive wheels, 101–102, 117–121,
        209–210, 211–212
    vs cutting tools, 100
Biflexing of coated abrasive belts,
        107
Billet conditioning (*See* rough grinding)

297